The Birth of Chemistry

"A nature is delighted by another nature, a nature conquers another nature, a nature dominates another nature."

Valentin Chirosca

The Birth of Chemistry

"A nature is delighted by another nature, a nature conquers another nature, a nature dominates another nature."

Valentin Chirosca

This book is for sale at https://www.createspace.com/4983646

This version was published on 2014-09-06

ISBN 978-1501063237

This is a Leanpub book. Leanpub empowers authors and publishers with the Lean Publishing process. Lean Publishing is the act of publishing an in-progress ebook using lightweight tools and many iterations to get reader feedback, pivot until you have the right book and build traction once you do.

©2014 Valentin Chirosca

But if any one should condescend to consider such sciences as are deemed rather curious than sound, and take a full view of the operations of the alchemists or magi, he will perhaps hesitate whether he ought rather to laugh or to weep. For the alchemist cherishes eternal hope, and when his labors succeed not, accuses his own mistakes, deeming, in his self-accusation, that he has not properly understood the words of art or of his authors; upon which he listens to tradition and vague whispers, or imagines there is some slight unsteadiness in the minute details of his practice, and then has recourse to an endless repetition of experiments: and in the meantime, when, in his casual experiments, he falls upon something in appearance new, or of some degree of utility, he consoles himself with such an earnest, and ostentatiously publishes them, keeping up his hope of the final result. Nor can it be denied that the alchemists have made several discoveries, and presented mankind with useful inventions. But we may well apply to them the fable of the old man, who bequeathed to his sons some gold buried in his garden, pretending not to know the exact spot, whereupon they worked diligently in digging the vineyard, and though they found no gold, the vintage was rendered more abundant by their labor.

Fr. Bacon, "Novum Organum." LXXXV

Contents

Foreword i

Contents. iii
    Illustrations iv

Dedication 1873 v

Chapter I. 1

Chapter II. 7

Chapter III. 15

Chapter IV. 21

Chapter V. 27

Chapter VI. 31

Chapter VII. 39

Chapter VIII 45

Chapter IX. 53

Chapter X. 61

Chapter XI 69

Lecture Notes 75
    Publication bibliography 80

Foreword

I have endeavored in the following pages to trace the rise and early development of a Very old Science, mainly that we may mark the attitude of thought which actuated the "scientific mind"; in bygone times, and may thus be led to compare the ancient with the modern method of evolving ideas, and building them up into a connected whole. With this object in view I have chosen the earlier history of the Science of Chemistry, in its various phases of: (a) primitive theories affecting the history of matter; (b) metallurgical chemistry of the ancients; (c) alchemy; (d) early ideas respecting the nature of combustion; and (e) the rise of pneumatic chemistry. The survey has been carried no farther than the time of the fathers of modern chemistry, Lavoisier, Priestley, Scheele, Bergman, Black, Cavendish, and Davy. The labors of these men belong to the later history of the Science.

G. F. RODWELL.

Marlborough,

Nov. 24th, 1873

Contents.

Chapter I

Introduction - Ancient Science - Origin of Chemistry - Definition of the Name - Definitions of Chemical Science

Chapter II.

Early ideas relative to the formation of the World - Thales of Miletus - Later Beliefs in his Doctrine - Anaximenes - Empedokles - Herakleitos - Anaxagoras - Demokritos - The Atomic Theory - Aristotle - The Ethereal Medium - Transmutation of the Elements - The Four-element Theory - Mode of interpreting it - Cause of the absence of Natural Science among the Ancients

Chapter III.

Practical Chemistry of the Ancients - Metallurgy: Gold, Silver, Electrum, Copper, Bronze, Tin

Chapter IV.

Iron, Lead, Quicksilver - Colors used for Painting and Dyeing - Glass - Certain Minerals known to the Ancients - Miscellaneous Processes

Chapter V.

Association of the seven Metals with the seven greater Heavenly Bodies - Consequent introduction of symbols into the history of Matter

Chapter VI.

The Alchemists - Origin of Alchemy - Hermes Trismegistus - Greek MSS. on Alchemy - Their probable Authorship and Age.

Chapter VII.

Latin and English MSS. on Alchemy - Sources from which the earlier Alchemists acquired knowledge - Arabic learning during the Middle Ages - Geber

Chapter VIII.

Avicenna - Albertus Magnus - S. Thomas Aquinas - Roger Bacon - Raymond Lulli - Arnoldus de Villa Nova - George Ripley - Basil Valentine

Chapter IX.

General Character of Alchemy and the Alchemists - The Pretiosa Margarita Novella - An Alchemistical Allegory - Alchemical Symbols - Paracelsus - Libavius

Chapter X.

Early Ideas concerning the Process of Combustion - Association of Nitre with the Air, so far as the part they play in Combustion is, concerned - Hooke's Theory of Combustion - Mayow's Experiments - Early Pneumatic Chemistry - Proof of the Analogy existing between Respiration and Combustion

Chapter XI.

The Theory of Phlogiston - Comparison with Hooke's Theory of Combustion - Early Ideas regarding Calcination - Stephen Hales - His Pneumatic Experiments - Boerhaave - Conclusion.

Notes

Illustrations

1. • Alchemical Representation of the Transmutation of the Elements
2. • Gold Washing: Fusion and Weighing of the Metal, from early Egyptian Tomb
3. • Furnace and Blow-pipe, from Egyptian Tomb
4. • Egyptian Bellows. Fifteenth Century B.c
5. • Smelting Furnace and Bellows used by Native Indians in the present day.
6. • Crux ansata of the Egyptians; Assyrian Symbol of Astarte; Later Symbol of the planet Venus
7. • Hermes Trismegistus, from the Temple at Pselcis
8. • An Alembic, and Symbols from Greek MSS (manuscripts) on Alchemy
9. • Alchemical MS (manuscript) of the Thirteenth Century. - British Museum
10. • English MS.(manuscript) on Alchemy - Fifteenth Century
11. • Distillation apparatus, from Geber's works
12. • An Alchemist hermetically Sealing a Flask containing a Solution of Gold
13. • Alchemical Representation of Processes
14. • Alchemical Representation of Processes
15. • Allegorical Representation of Transmutation
16. • Allegorical Representation of Transmutation
17. • Symbols of Lead, from Italian MS. (manuscript) of the Seventeenth Century
18. • Designs from Mangetus (Bibliotheca Chemika Curiosa)
19. • John Mayow from his "Tractatus Quinque Medico-Physici, 1674."
20. • Early Experiment in Pneumatic Chemistry
21. • Early Experiment in Physiological Chemistry
22. • Hale's Method of Measuring a Gas
23. • Measurement of the Elastic Force of the Gas produced by Fermenting Peas
24. • Hale's Pneumatic Experiments

Dedication 1873

Quod si quis ad intuendum ea, qua magis curiosa habent quam sana, animum sub miserit, et alchymistarum aut magorum opera penitus introduxerit, is dubitabit forsan, utrum risu, an lacrymis potius, illa digna sint. Alchymista enim spem alit aeternam, atque ubi res non succedit, errores proprios remos substituit; secum accusatione reputando, se aut artis aut authorum vocabula non satis intellexisse, unde ad traditiones et auriculares susurros animum applicat aut in practica sua scrupulis et momentis aliquid titubatum esse; unde experimenta in infinitum repetit; ac interim quum inter experimentorum sortes, in quadam incidit aut ipsa facie nova, aut utilitate non contemnenda; huiusmodi pignoribus animum poscit, eaque in maius negandum et celebrat; reliqua spe sustentat. Neque tamen negandum est alchymistas non pauca invenisse, et inventis utilibus homines donasse.

Fr. Bacon, " Novum Organum." LXXXV

Chapter I.

Introduction - Ancient Science - Origin of Chemistry - Derivation of the Name - Definitions of Chemical Science.

The history of a natural science resembles in many respects the history of a nation. In each instance the object is first to obtain a knowledge of causes, then to frame laws. The first are those causes which most promote the well-being of the nation, the second those causes which produce the phenomena of the Universe. In each instance we start with an absence of all law, and we may observe the slow efforts of the human mind to trace each effect to its proper cause, to group together causes, and finally to connect them by one bond. The main difference is this, that in the case of the nation man has to deal with laws which must be founded upon a just study and close observance of every phase of that particular community, influenced as it is by numberless external causes, such as race, climate, religion, habit of thought, tradition; while in the case of the science he has to evolve preexistent laws, also by the close observance of facts, which are hidden from him by the complex mechanism of nature. M. Taine would tell us that the laws which influence the development of peoples are just as absolute, definite, and preexistent, as those which govern the affairs of nature; but we are quite disinclined to admit this, even in regard to one particular race, in one particular locality. In both histories we have similar forms of government, similar assemblies of lawgivers; we have our aristocracies, oligarchies, democracies, republics: we have at some period or other Conservatives and Liberals of every shade. We know not what Conservative rule can compare with the dominance of the science of Aristotle for twenty centuries, and we cannot be too ready to welcome the Liberal-conservative era of Copernicus and Giordano Bruno, the Liberal era commenced by Galileo and Francis Bacon, which by easy stages is passing, if it has not passed, into the right Radical era of modern scientific thought. The "Republic of learning" is no empty phrase.

No one would venture to deny the value of a knowledge of the history of nations, and we are inclined to believe that the history of the natural sciences is not without its uses. It is neglected because during the last century new discoveries have quickly succeeded each other, old sciences have augmented, while new sciences have arisen; in fact, the progress of science has been so extraordinarily rapid that we have scarcely time to turn aside and look at its past history; the present is sufficient for us, and if we once get out of the main current of thought we have difficulty in regaining lost ground. Yet we may no more forget that we owe our present wise laws and great constitutional system to the labors of ten centuries of men, than that our science of today represents the accumulation of the scientific thought of twice ten centuries. Intellectual revolutions have not been less frequent than social revolutions, nor battles of the pen than battles of the sword; the crash of a fallen philosophy has often been louder than that of a fallen throne; the wail of the last Phlogistians rent the heavens; the Aristotelian physics died with groaning, and gapings, and a discolored visage.

In tracing the history of a science, we are first led to inquire whether the ancients possessed any knowledge of it, and whether it originated among them. Now, the ancients made but little progress in any of the natural sciences. They divided all human knowledge into three parts: Logic, or mental philosophy; Physics, or natural philosophy; Ethics, or moral philosophy. Some placed logic first, some ethics, but no one physics. Philosophy was compared to an egg: logic the shell, physics the white, ethics the yolk; or, again, it was compared to a living creature: logic the bones, physics the flesh, ethics the soul. Plato separates logic as the knowledge of the immutable, from physics the knowledge of the mutable. The Cynics sought a complete freedom from any object or aim in life, and renounced all science. Socrates aimed at logical definition, and affirmed that the true nature of external objects can be discovered by thought without observation. The knowledge of ones self (γθω σεαντον, αυτογνωσίας) is the true object in of all philosophy. Others of the ancients that geometry might be employed for the element of land, and astronomy cultivated so: might be of use to sailors, but on no account as serious subjects of mental occupation. Know-obtained from external sources is worthless; there is nothing to be learned from fields and trees, A certain philosopher is said on this principle to have eyes out, in order that his mind might not be influenced by external objects, and might be left to pure contemplation. (How curiously this contrasts with plaint of Galileo just before his death: "Proh dolor the sight of my right eye, that eye whose, I dare say it, have had such glorious results is forever lost. That of the left, which was and is imperfect, is rendered null by a continual weeping.")

Thus it happened that natural science made but progress among the ancients; thus it happens schoolboy of twelve knows more about earth,and fire, and water, than was dream of in the philosophic of the greatest thinkers of antiquity. Let us however, give them their due; let us confess that Plato possessed the " finest of human intellects, exercising boundless dominion over the finest of languages;" that Aristotle was the greatest genius the world has ever seen; that as pure intellectual evolution they have handed down to us a mass of grand philosophy; ten thousand noble efforts human spirit. Everything favored the exercise of the unaided intellect, while it is hard to the difficulties which presented themselves investigation of nature. At one period it was considered impious to attempt to explain the manifestations of the gods. There was an outcry in Athens, a popular demonstration, when the thunderbolts of Zeus were referred to common fire produced by the .collision of clouds. The feeling was of the same nature as that conveyed by Campbell's Stanza:

"When Science from Creation's face

Enchantment's veil withdraws,

What lovely visions yield their place

To cold material laws !"

Only the feeling existed in an intensified form, for here the first of the gods was derided the Olympian Zeus, Lord of the Air, he who rides upon the storm, and hurls the thunderbolt. For a length of time, therefore, any investigation of nature was impossible for religious reasons. Men were to worship nature, to be filled with awe and wonder— δεισιδαιμονια - superstition - in presence of great natural phenomena, but not to inquire too closely into their causes. Twenty centuries later the Doctors of Salamanca who interrogated Columbus, the Inquisitors of the Sacred College who examined Galileo, upheld the same old doctrines, albeit the old gods had passed away. But the investigation of nature was impossible among the Greeks; their capabilities were very limited, they had no instruments for observations, or experiments of any kind, neither had they the faculty of observation; their minds were untutored in that particular direction. Then they had to contend against their own particular habit of thought, the extreme tendency to concretion, to hasty generalization from mental premises; or if an observation had made, a broad general law was deduced from it to further observation. So also the Chaldean's and Parsis had to contend against the mysticism, the astrology, and magic, which originated among them; and the ancient Hindu was so given to extreme abstraction, and to the evolution of all manner of strange metaphysical dogmas, that we could scarcely or much science from an Eastern source, Egyptian learning was monopolized by the priests, and they so wove together the real and the unreal, and were so secret withal in their actions, that although much of the Greek learning came direct Egypt, we cannot trace it to its direct source, it to one Egyptian writer on philosophy. The Greeks, too, received much from the Phoenicians; but here also we find no record. We will presently inquire more fully into the exact amount of science possessed by the ancients.

We have chosen for our historical survey one of oldest of the natural sciences, for obvious reasons, the chief one being that it will enable us to observe minutely the early thought of ancient peoples in regard to certain phenomena of nature. The science of Chemistry does not owe its existence to any one people, or to any sudden process of development The time when the foundation-stone was laid is too remote to be even suggested; the basis of the edifice is sunk deep in Eastern soil; the walls were slowly and laboriously raised during the Middle Ages, and were completed by Lavoisier, Blank and Priestley; the men of our day are working at the roof. We neither hold with M. Goguet that Moses possessed considerable knowledge of chemistry because he dissolved the golden calf, nor with M. Wurtz, when he says "La chimie est une science Franchise. Elle fut institute par Lavoisier d'immortelle memoire." Chemistry was not a science until long after the time of Moses; it was a science long before the time of Lavoisier. We wonder what Dr. Hermann Boerhaave of Leyden (whose large quarto "Elementae Chemise" was published in 1732, nine years before the birth of Lavoisier; would say to the proposition of M. Wurtz. Short of this, it would be difficult to overrate the services which Lavoisier rendered to chemistry. But the science has grown up by a gradual process of evolution; upon its surface we find the impress of many and diverse phases of thought and of action; the science of today is the summation of many intellectual efforts produced by the constant struggle of the human mind for truth. How often that truth has been hidden by a mass of sophistries; how often it has been absorbed by some false philosophy to appear again untarnished in due time; how often the attempt has been made to crush it under foot; and how it has ever risen to the surface at last, all who read the history of faiths, nations, ideas, must know. It will be our object to show this is the study of the particular science which now engages our attention.

The word "χημεία - chemistry" first occurs in the Lexicon of Suidas, a Greek writer of the eleventh century; he defines it as "the preparation of gold and silver." In the "Lexicon Graeco-Latinum " of Robertus Constantinus, published in 1592, the same definition is given, and Suidas is quoted as the authority. According to Olaus Borrichius, however, there were Greek writers on alchemy before this date; there is said to be a Greek MS. of the fifth century on alchemy in the King's Library in Paris, and others of a somewhat later date in the libraries of Munich, Milan, Venice, Hamburg, and Madrid; but we are inclined to doubt whether any of these were written before the ninth or tenth century. They are probably the work of monks living at Alexandria and Constantinople; indeed, one of them is entitled, "Cosma the Monk, his Interpretation of the Art of making Gold." The titles of some of the others will prove to us that we can place but little faith on any date which may be assigned to them:

- "Heliodorus on the Art of making Gold" Ἡλιοδώρου φιλοσόφου πρὸς Φεοδόσιον τὸν μέλαν Ἡliodώroy filosófoy pròs Theodosíoy tòn melánh Iliodorou philosopher TO Tieodosiou the Black (i. e. Alchymy), δἰ Ἰάμβων

    It exists in MS. in several libraries in Europe, and is printed, from the Paris MS., in Fabric. Bibl. Graec. vol. viii. p. 119. The title is Ἡλιοδώρου φιλοσόφου πρὸς Φεοδόσιον τὸν μέλαν *Basile/a, peri\ th=s tw=n filoso/fwn *Mustikh=s te/xnhs (i. e. Alchymy), δἰ Ἰάμβων Kühn and Hoffmann (Lex. Bibl. s. v.) believe the poem to be genuine, but Jacobs calls it the clumsy fabrication of a later time, to which the name of Theodosius was prefixed to give it the semblance of authority; and he suggests that the name Heliodorus may have been used, after the fashion of the Alchymists and Rosicrucians, on account of its etymological signification.

- "John the High Priest, in the Holy City, concerning the Holy Art."
- "Isis the Prophetess to her son Orus."
- "Moses the Prophet on Chemical Composition" περίπου Χημική Σύνθεση.
- "Cleopatra on the Art of making Gold." - or Chrysopoeia of Cleopatra, Black and white represent the duality present in everything and the Greek inscription in the center is εν το παν (hen to pan) "all is one";
- "Democritus the Abderite, the Natural Philosopher, on the Tincture of Gold and Silver, and on Precious Stones and Purple."

Equally worthless, we believe, are the Greek derivations of the word "chemistry". Many (among others M. Hoefer) derive the word from λυώνω, to fuse or melt, because the majority of old chemical operations were effected by fire witness calcination, ignition, distillation, sublimation, desiccation, reverberation. The earliest chemical arts, such as the smelting of metals and the production of glass, were also operations of «fire. Indeed, the science has been called Pyrotechnia (Πυρ τεχνή fire Art, the art of fire), because, says Lemery, in his "Cours de Chimie", "we in effect produce all chemical operations by means of fire." Others derive chemistry from χεω —that which is poured out, a liquid, in allusion to the various liquids used in chemical operations; but this derivation is not worth a moment's notice. We must rather look to an Egyptian source Plutarch tells us that Egypt was called Chemia, on account of the black color of the soil, and that the same term was applied to the black of the eye, which symbolizes that which is obscure and hidden. This word is related to the Coptic khems or chetns, which also signifies obscure, occult, and is connected with the Arabic chema, to hide. It is probable that we have here the true derivation of the word chemistry. The first treatise on the science, the date of which is known with any certainty, was written by the Arabian Yeber or Geber, and at that time (the eighth century), Arabic learning had considerable influence on European culture. The science was called the occult, or hidden, because it related principally to the secret art of the transmutation of metals, as the definition of Suidas, given above, and the earlier works on the science, prove. The term black art has been applied both to alchemy and to the magical arts so often associated with it, and clearly agrees with the above derivation. The al in alchemy is the Arabic particle the, so that alchemy signifies "the hidden science" par excellence; we notice the same prefix in #/koran, alcohol (the burning liquid), alkali (the acrid substance), #/gebra, alembic (the cup-shaped vessel), and in the names of many stars, as Aldebaran, Algenib, Alpheratz,—all words of Arabic origin.

Whatever difficulties there may be in determining the precise derivation of the word chemistry, there can be none in defining the science as distinctly and definitely the science which treats of the changes which matter undergoes; while physics proper treats of the action of various forces: heat, light, electricity, magnetism upon matter, in all cases unaccompanied by any change of composition. If we heat a piece of iron to redness, or cause it to convey an electric current, or place it in contact with a magnet, it has been submitted to various actions, but when they are removed it returns to its original condition. On the contrary, if we fuse it with sulphur a chemical change takes place, a new substance is formed, and the iron does not return to its original condition. This idea of change is the fundamental chemical conception. The first man who made glass, or extracted a metal from its ore, effected a chemical change; the idea became most sovereign and dominant in alchemy, the attempt to change base metals into gold; it reigned throughout the period of phlogistic chemistry, for was not phlogiston a subtle entity which effected changes in matter according as it was assimilated by matter or rejected from it? It is equally the character of the chemistry of Lavoisier and Cavendish, of Davy and Dalton, of Berthelot and Cannizzaro.

The "philosopher's stone" (of which much more anon) was a substance supposed to change all things into gold; the "elixir vitae" was a substance which was to change old men into youths; the "universal solvent" was to change everything to a liquid form. Let us look at some of the definitions of chemistry. Boerhaave says: "Chemistry is an art which teaches the manner of performing certain physical operations, whereby bodies cognizable to the senses, or capable of being rendered cognizable, and of being contained in vessels, are so changed by means of proper instruments, as to produce certain determined effects, and at the same time discover the causes thereof, for the service of various arts." Sir Humphrey Davy writes as follows: " Most of the substances belonging to our globe are constantly undergoing alterations in sensible quantities, and one variety of matter becomes, as it were, transmuted into another. Such changes, whether natural or artificial, whether slowly or rapidly performed, are called chemical; thus the gradual and almost imperceptible decay of the leaves and branches of a fallen tree exposed to the atmosphere, and the rapid combustion of wood in our fires, are both chemical operations. The object of chemical philosophy is to ascertain the causes of all phenomena of this kind, and to discover the laws by which they are governed." Quite recently Dr. Miller has defined chemistry as " the science which teaches us the composition of bodies," and such knowledge we can only obtain by pulling matter to pieces (analysis), or by building it up (synthesis). Dr. Hofmann of Berlin has defined the vast body of so-called organic chemistry as "Cosma the Monk, his Interpretation of the Art of making Gold. the history of the migrations of carbon", and is not migration change of place?

Let us then define chemistry as the science which treats of the various kinds of matter, whether simple or compound, of which the world is composed, their properties, and the laws which govern their combination with, and separation from, each other.

We will now discuss any ideas of the ancients which bear upon changed matter in any form or condition: thus, their early cosmogonies; the knowledge they possessed of metals and compound bodies; and their various technical operations, such as glass-making and smelting, alike demand our attention.

Chapter II.

Early ideas relative to the formation of the World - Thales of Miletus: Later beliefs in his doctrine - Anaximenes - Empedokles - Herakleitos - Anaxagoras - Demokritos - The Atomic Theory - Aristotle - The Ethereal Medium - Transmutation of the Elements - The Four element Theory - Mode of interpreting it - Cause of the absence of Natural Science among the Ancients.

If we compare all the earliest ideas as to the formation of the world, we find the resolve themselves into the belief that the ether and chaos, mind and matter, were the original principles of things. The ether, a subtle vivifying principle, "passing as a mighty breath over the chaos; the chaos a boundless watery expanse without form." It was thus according to Sanchuniathon in the belief of the Phoenicians, and the twenty-five principles of the Hindu philosophy of San'chya are finally reduced to these matter and spirit, nature and soul. The Egyptian deity was called Num as the spirit moving over the face of the waters, Pthah as the principle of production. The Hindu deity Brahme typified the productive force of nature. Among more western nations, Gaia, the personification of earth, was held to be the first that sprung from Chaos, and the wife of Ouranos.

Okeanos was their son, and according to Homer was source of all the gods. The worship of the elements, and of the sun and moon, was among the very earliest forms of worship; thus we have in India, Agni the god of fire, Indra the god of the firmament; the sun was sometimes worshiped as a symbol of the deity, sometimes as a deity; fire was worshiped by the ancient Persians as a symbol of the deity; in the Homeric religion we find the Olympian Zeus, the lord of the air, who possesses absolute and universal power. We must notice, too, Aldoneus, the brother and lord of the Underworld, said by some of the Greek philosophers to designate earth, and undoubtedly an old nature power. Again, "Hephaistos" says Mr. Gladstone, " bears in Homer the double stamp of a Nature Power, representing the element of fire, and of an anthropomorphic deity who is the god of art at a period when the only fine art known was in works of metal produced by the aid of the fire is also one of the seven star deities of Chaldaea, the symbols and names of which were given at an early date to the seven metals."

The elements of the Greek philosophers were, as we shall presently show, rather principles than elements in the sense in which we speak of the sixty-five elements now known to chemistry. There was a marked tendency in the earliest period of Greek philosophy to make one element or principle fundamental, and to evolve the other elements and the world from it. Thales of Miletus, who lived in the sixth century B.C., and who was called "the first of natural philosophers" by Tertullian, and "the first who inquired after natural causes" by Lactantius, affirmed that water was the first principle of things, perhaps, according to some writers, because Homer had made Okeanos the source of the gods. At least we are reminded of the boundless watery chaos of older cosmogonies. This doctrine of Thales was not without its supporters during the Middle Ages, and, indeed, the convertibility of water into earth and air was not absolutely disproved until about a century ago. One of the ablest supporters of the dogma was Jan Baptista Van Helmont (b. 1577, d. 1644), who affirmed that all metals, and even rocks, may be resolved into water; animal substances are produced from it, because fish live upon it; and vegetable substances may be also produced from it This last assertion he endeavored to prove by what would appear to be a very conclusive experiment in those days, when neither the composition of the air, nor of water was known. He took a willow which weighed five pounds, and planted it in two hundred pounds of earth, which he had previously carefully dried in an oven. The willow was frequently watered, and at the end of five years he pulled it up and found that its weight amounted to one hundred and sixty-nine pounds and three ounces. The earth was again dried, and was found to have lost only two ounces. Thus it appeared that 164 lbs. of wood, bark, roots, leaves, &c, had been produced from water alone. Hence he inferred that all vegetables are produced from water alone; not knowing, as was afterwards proved by Priestley, that a constituent of the atmosphere called carbonic acid had furnished the solid part of the tree, although, indeed, there was much water with it. Boerhaave devotes a page of his big book to a discussion of "whether water be convertible into earth." He concludes that the small earthy deposit observed rain-water is distilled, arises from the particles it which had settled on the water before its introduction into the distilling vessel. Mr. Boyle had previously affirmed that "a very ingenious person, who had tried various experiments on rain-water, put him beyond all doubt about this transmutation, for he solemnly affirmed, on experience, that rain-water even after distillation in very clean glasses near two hundred times, afforded him this white earth.' Finally, Lavoisier, in 1770, communicated to the Academie des Sciences an elaborate paper, "on the of water, and the experiments by which it has been attempted to prove the possibility of changing in to the earth." In this he conclusively proved that water cannot be changed into earth, although it be distilled backwards and forwards for many successive dais. Here then we find the old Thalesian theory at the last disproved, but not before it had endured for twenty-four centuries; and this is by no means a solitary example of the permanence of old ideas. We shall become acquainted with yet older theories, are still admitted, and which seem to be essential to physical philosophy.

On the other hand, Anaximenes regarded air as the primal element, Herakleitos fire, Pherekides earth and some philosophers grouped two elements together. Anaximenes held that clouds are caused condensation of air, rain by the condensation of clouds; he appears to have clearly connected condensation with cold, rarefaction with heat. Archelaus affirmed that air when rarefied becomes fire, or when condensed, water. It was very generally believed during the Middle Ages that water when boiled was converted into air. Empedokles introduced the idea of four distinct elements—earth, air, fire, and water, not capable of passing one into the other, but forming all things by their inter mixture. These elements are acted upon by two principles, a uniting force of amity, a separating force of discord, corresponding somewhat to our attraction and repulsion. He endeavored to prove the four-element theory by the following crude experiment: wood is burnt upon a hearth, fire seems to be evolved from it, the smoke is air, moisture is deposited on the hearthstone, while the ashes are earth:—hence wood is made up of earth, air, fire, and water. Empedokles was one of the first to materialize the Homeric gods; he applied his four element theory even to them, declaring that Zeus was the element of fire, Here the element of air, Nestis the element of water, and Aidoneus the element of earth. Herakleitos (about 460 B.C.) made fire the primal element, and assumed that it condensed itself into the material elements, and that air, water, and earth were respectively formed as the fire became more condensed. He asserted, moreover, that all things are in perpetual motion and change, the moving force being fire; "fire is to him," says Schwegler," even in individual things, the principle of movement, of physical as of spiritual vitality; the soul itself is a fiery vapor. "We find in the fire of Herakleitos to some extent the attributes of what we now call a physical force; thus it is precedent to matter, and is the motive power of the universe; it influences and changes matter; it is perpetually undergoing transformation, but ultimately returns to its own form. Prof. Max Mtiller speaks of Herakleitos as "one of the boldest thinkers of ancient Greece". We can well understand why fire should, at a very early date, be regarded as chief of the elements, and the motive power of the universe; it had long been worshiped as a symbol of the deity by the Chaldaeans; a worship which possibly originated with the Scyths; for Zoroaster, who introduced fire-worship among the Medo-Persic races, is supposed to have been a Scythian. Again, Agni, the god of light and fire, was placed first in the Hindu Trinity.

Anaxagoras of Klazomene (500 B.C.) asserted that, originally all things existed in infinite disorder; before the Creation there was a chaos of mingled particles of matter, which were arranged in order by a designing intelligence, or mover of matter. The primitive constituents of things are not definite elements, like those of Empedokles, but are homceomeries that is like parts, small particles of matter like the masses they produce when they aggregate. Thus a mass of iron is produced by the aggregation of an infinite number of iron-homceomeries brought out of the chaos by the νοῦς (mind), which latter possesses vortical motion which enables it to separate like parts and bring them together, somewhat on the principle of gold-washing.

If a dish containing substances of different relative weight, such as cork, sand, and lead shot, intimately mixed together, be caused to rotate, like particles" will come together, the lead in one place, the sand in another, and this experiment will help us to realize to some extent the meaning of Anaxagoras when he assumes that the vortical motion caused homceomeries to aggregate and form the world. Leukippos taught that the world is produced by the falling together of small indivisible particles or atoms Greek ἄτομος (atomos, "indivisible") from ἀ- (a-, "not") and τέμνω (temnō, "I cut"), which are the principles of things, and which possess rapid circular motion. Demokritos (460 B.C.) extended the atomic theory of Leukippos; he contended that the principles of things are atoms and a vacuum. The atoms are invisible by reason of their smallness, indivisible by reason of their solidity, impenetrable and unalterable. They have no other qualities, neither heat, nor cold, nor color. Atoms are infinite in number, the vacuum is infinite in magnitude. Atoms differ from each other in size, shape, and weight. They are actuated by necessity or fate αναγκαιότη, and possess an oblique motion in the vacuum which causes atoms of like shape to collide, and group themselves together, by which means all things are formed. The vacuum is necessary, otherwise motion of the atoms would be impossible, because there would be (no place to receive them. Long before the time of Demokritos an atomic theory had been proposed in India by Kanada, the founder of the Nyaya system of philosophy, of which this theory forms the distinguishing feature. The theory of Leukippos is quoted by Possidonius to Moschus, a Phoenician.

During the Middle Ages many writers made the atomic theory a prominent part of their system. Descartes adopted it in a somewhat modified form, associated with his particles the vortical motion possessed by the homceomeries of Anaxagoras. Finally almost in our own day, the atomic theory introduced into chemistry by Dalton, and its introduction marked an important era in the science. At the present time the doctrine of atoms forms principal feature in chemistry, and other branches of science find the conception most conducive to philosophical explanation of phenomena. The definition of an atom given by Demokritos is almost as absolute and precise as that which we in our most modern treatises. Thus the theory endured for more than twenty-five centuries, is likely to endure until there shall be no science. It offers a striking example of oneness of physical thought; the conception is to be essential to Natural Philosophy; the stupendous phenomena may be referred to lie motions. S. Augustine has well said, "Deus lagnus in magnis, maximus autem in minimis."

The Hindus not only possessed the idea of the lie constitution of matter, but further associated attractive force with the atoms. This is well shown in the following extract given by Sir William Jones, from the poem of "Shi'ri'n and Ferhad, or Divine Spirit, and a human soul disinterestedly pious": There is a strong propensity, which dances through every atom, and attracts the minutest particle to some peculiar object; search this Universe from its base to its summit, from fire to air, from water to earth, from all below the moon to all above the celestial spheres, and thou wilt not find a corpuscle destitute of that natural attractibility; the very point of the first thread in this apparently tangled skein is no other than such a principle of attraction, and all principles beside are void of a real basis; from such a propensity arises every motion perceived in heavenly or in terrestrial bodies; it is a disposition to be attracted, which taught hard steel to rush from its place and rivet itself on the magnet; it is the same disposition which impels the light straw to attach itself firmly to the amber; it is this quality which gives every substance in nature a tendency towards another, and an inclination forcibly directed to a determinate point."

The most prolific writer on Science among the ancients was Aristotle (b. 385, d. 322 B.C.). He was the author of various treatises, on the Heavens, on Generation and Corruption, on Physics, on Respiration, on Audibles, &c, and his views as well on metaphysics and ethics, as on science, were nearly universally accepted during the Middle Ages. Indeed, the scientific writings of Aristotle influenced science for nearly twenty centuries. Few, however, join of his opinions concern us here. He was the first to introduce into Greek philosophy the ether, which he regarded as a fifth element (hence afterwards called quinta essentia) more subtle and divine than the other elements. The word quintessence is frequently used by the alchemists and early chemists and is found in our most recent English dictionaries.

The idea of an infinitely rarefied and all-penetrating matter had long existed in physical philosophy, notably in the Hindu systems; it was probably recognized as a fifth element prior to the ninth century B.C. Aristotle is said to have called it "αιθηρ - ether" from αει - universities and θεω - considered, because he conceived it to be always in motion, and to be the moving agency of the other elements; but we cannot admit this derivation now. In the present day we find it impossible to explain various phenomena, notably those connected with radiant heat and the polarization of light, without assuming the existence of some rare ethereal medium, cubic miles of which would not weigh a milligram, and we still call it the ether. Few physical systems have avoided this supposition; we make less use of it in chemistry than in physics; but it would be difficult to account for such actions as the combination of chlorine and hydrogen under the influence of light, without it.

Aristotle held that the four elements are mutually convertible, and he assigned two qualities to each, one of which was common to some other element, Thus he said:

Fire is hot and dry.

Air is hot and moist.

Water is cold and moist.

Earth is cold and dry.

In each of these one quality is dominant. Thus fire is more hot than dry, air more moist than hot, water more cold than moist, and earth more dry than cold. If the dry of fire be vanquished by the moist of water, air will result; if the hot of air be vanquished by the cold of earth, water will result; if the moist of water be vanquished by the dry of fire, earth will result. This idea of the transmutation of the elements was adopted generally in works on alchemy; the above figure, which embodies it, is from a work entitled "Preciosa Margarita Novella", published in Venice in 1546.

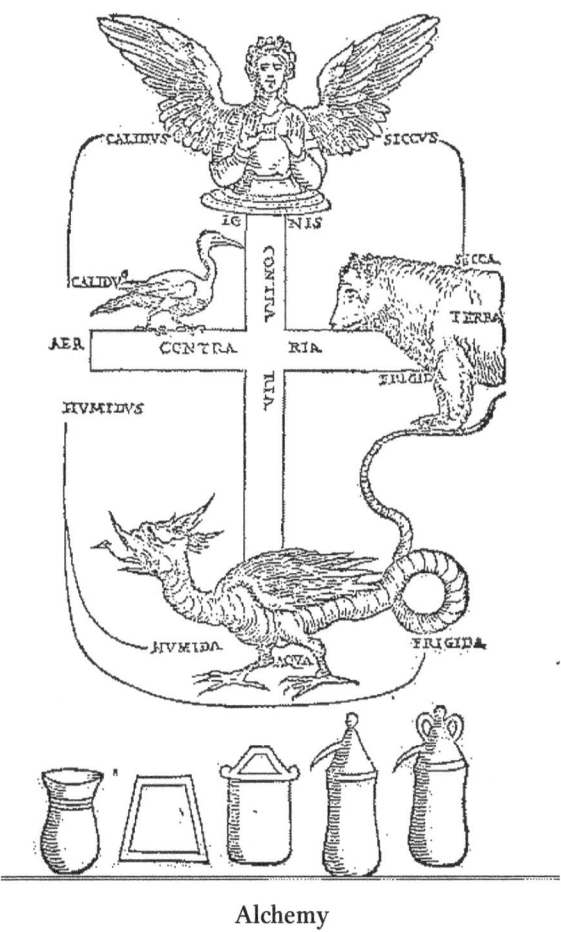

Alchemy

Fig. 1 Alchemical Representation of the Transmutation of the Elements.

Aristotle's method of expressing the transmutation of the elements does not seem to differ much from that of earlier philosophers; it would appear that he means to imply that if water be heated air is produced, while if it be heated more strongly so as to evaporate it to dryness, earth is left. His account of the generation of fire from air and earth is based on the most shallow and meager observation, and shows to what results the most astute mind may be led if unaided by experiment.

The generation of fire, he says, is made evident by the senses, for flame is notably fire, but flame is burning smoke, and smoke is from air and earth.

It is not here that we may tell how the philosophy of Aristotle was introduced into Europe by the Arabians, how from it arose that stupendous mass of false philosophy and perverted Aristotelianism called Scholasticism, and how for centuries the blind acceptance of the Peripatetic dogmas retarded the progress of science. Worse than all, Averroes, who has been called "l'ame d'Aristote", and who scattered Aristotelianism broadcast over Europe, did not know Greek, and the Latin versions of Averroes were "Latin translations from an Hebrew version of an Arabic commentary on an Arabic translation of a Syriac version of a Greek text." We may not, therefore, blame Aristotle for the results which followed from the too general and literal acceptance of his philosophy. Mr. Lewes has well said, "However he may have been impelled to systematize on imperfect bases, and to reason where he should have observed, it is not too much to say that had he reappeared among later generations, he would have been the first to repudiate the servility of his followers, the first to point out the inanity of Scholasticism. His mighty and eminently inquiring intellect would have been the first to welcome and to extend the new discoveries. He would have sided with Galileo and Bacon against the Aristotelians."

We have spoken above of the endurance of the Thalesian theory, that all things are formed from water, and of the yet older theories of the existence of an ethereal medium, and of atoms; but the theory which affirms that the world is composed of the four elements—earth, air, fire, and water, is yet older, and is, indeed, the oldest physical theory of which we have any knowledge. It certainly existed before the fifteenth century B.C., it was adopted in India, Egypt, and, as we have seen, in Greece at a very early date. Then in the case of those philosophers who made water, air, fire, &c, primal elements, this element was first transmuted into the three other elements, and the world was formed from the four. We must be careful, however, to remember that these four elements are not to be understood too literally, they were rather principles or types of qualities than actual elements. Several philosophers divided fire into a purer and grosser part. Seneca tells us that the Egyptians extended the theory by assigning to each element an active and a passive form: thus fire was divided into light which shines, and into fire; air into passive atmosphere and active wind; water into fresh and salt, water; and earth into cultivating land on the one hand, and rocks on another. These elements were extended yet more, In later times Fire would come to signify every-appertaining to ignition; thus light, whether accompanied by heat or otherwise, flame, the heat it in all bodies, incandescent bodies, stars, fiery meteors, lightning, and all visible manifestations of electricity, would be included under the term. Air would include smoke, steam, all vapors, and whatsoever approached to the nature of a gas. When were first discovered a hundred years ago, they where called Airs; thus we read of fixed air, nitrous air, dephlogisticated air, &c. Water would: all liquids, of which, no doubt, blood, milk, wine and oil, were in early times the most familiar; the words aqua fortis, aqua regia, aguardiente, eau-de-vie, are vestiges of the old practice. Earth included all rocks, however dissimilar they might be, all kinds of cultivable land, metals, and whatever appertained to solidity. Every solid was regarded as a kind of earth at first. A century ago many ices were called earths. At the present time the sixty-five elements known to the chemist, eight are classed as "earths" and three as "alkaline " The fact is, the four ancient elements , were types of great classes of which the whole world constituted.

In their most general sense, earth, air, signified solidity, liquidity, gaseous, while fire was the force exercising itself upon matter. We have seen that the elemental fire of Herakleitos mover of matter, the principle of movement, which produces perpetual changes around us. Fire was the ψυχή, the anima, the soul, the vivifying spirit. The mythological side of the belief is seen in the story of Prometheus, who is fabled to have stolen fire from Heaven and therewith vivified mankind. The philosophical side of the belief is seen in the dogmas of Herakleitos. The four-element theory evolved itself from the rude ideas about ether and chaos, mind and matter, before discussed; it is one of those crude physical theories which is enunciated and accepted by races the most diverse in character, country, faith, destiny. There is great oneness in the human mind in the matter of broad principles in crude cosmically ideas. And let us not forget that the four-element theory was universally accepted during the Middle Ages, and was only disproved a century ago, when air was proved to be a mixture of two gases, water a combination of two gases, fire the result of intense chemical action, and earth a mixture of some dozens of elementary bodies, some combined, some single. We do not deny that during the continuance of the four-element theory it may often have been taken in its strictly literal sense; but we do venture to assert that the richer and more cultured intellects regarded it in the light we have above described.

We can quite understand why there was so little natural science among the ancients, when we remember the absence of all experimental method and means, and the obstacle presented by the habit of mind which induced them to apply reasoning in place of experiment in the study of nature, to reason upon an immature or ill-observed fact, and to generalize upon altogether insufficient data. A simple sophistry applied to observation could lead to monstrous results. Take, for example, the argument of Diodorus, as given by Sextus to prove that nothing is moved:—"If a thing be moved, it is either moved in the place where it is, or in the place where it is not. But not in that wherein it is, because it rests in the place wherein it is; neither in that wherein it is not, for where is a thing is not, it can neither act nor suffer. Therefore nothing is moved". Again, Sokrates and many of his followers taught that it was unwise to leave those affairs which directly concern man, to study those which are beyond his control and external to him. Thus, to inquire into the nature and distance of the stars seems an useless speculation, because even if we could ascertain these things, neither alter the course of the stars nor to any benefit of mankind.

We have however, seen above that many of the Greek philosophers had more or less definite notions concerning matter and force, and that they frequently insist upon the transmutation of matter from one form into another; so far and so far only are we concerned with their dogmas in our inquiry into the Birth of Chemistry. But we must not fail to notice the existence at a very early date of the four-element theory, of an atomic theory, of the idea of a etheral medium, of the idea of transforming one kind of matter into another by the agency of some motive principle. Neither let us forget to note the similarity of principles in diverse philosophies; thus the homœomeries of Anaxagoras and the atoms of Leukippos are clearly related, so, too, are the νους nons of Anaxagoras, of Demokritos, the actuating form of fire of Herakleitos, the moving ether of Aristotle. The links which bind together ancient and modern physical thought are strong and enduring; and, since they have lasted during the rise and fall of many nations, and during the most profound changes in the mode and tone of thought, it is not unlikely that they will endure as long as the chain itself.

According to Anaxagoras the cosmos is made of infinitely divisible matter, every bit of which can inherently become anything, except Mind (nous), which is also matter, but which can only be found separated from this general mixture, or else mixed into living things, or in other words in the Greek terminology of the time, things with a soul (psuchē). Anaxagoras wrote: All other things partake in a portion of everything, while nous is infinite and self-ruled, and is mixed with nothing, but is alone, itself by itself. For if it were not by itself, but were mixed with anything else, it would partake in all things if it were mixed with any; for in everything there is a portion of everything, as has been said by me in what goes before, and the things mixed with it would hinder it, so that it would have power over nothing in the same way that it has now being alone by itself. For it is the thinnest of all things and the purest, and it has all knowledge about everything and the greatest strength; and nous has power over all things, both greater and smaller, that have soul [psuchē].

In supposing that void exists, the atomists deliberately embraced an apparent contradiction, claiming that "what is not" exists. Apparently addressing an argument by Melissus, a follower of Parmenides, the atomists paired the term for "nothing" with what it negates, "thing," and claimed that in a phrase typical of the atomists the one "no more" exists than the other (DK 67A6). Schofield (2002) argues that this particular phrase originated with Democritus and not his teacher Leucippus. By putting the full (or solid) and the void ontological on a par, the atomists were apparently denying the impossibility of void. Void they considered to be a necessary condition for local motion: if there were no unoccupied places, where could bodies move into? Melissus had argued from the logical impossibility of void to the impossibility of motion; the atomists apparently reasoned in reverse, arguing from the fact that motion exists to the necessity for void space to exist (DK 67A7). It has been suggested that Democritus conception of void is that of the (temporarily) unfilled regions between atoms rather than a concept of absolute space (Sedley 1982). Void does not impede the motion of atoms because its essential quality is that of "yielding," in contrast to the mutual resistance of atoms. Later atomist accounts attest that this "yielding" explains the tendency of bodies to drift into emptier spaces, driven out by collision from more densely packed regions (Lucretius DRN 6.906–1089).

Chapter III.

Practical Chemistry of the Ancients - Metallurgy: Gold, Silver, Electrum, Copper, Bronze, Tin.

In the preceding chapters we have discussed such theories of the ancients as involve the conception of change of matter (notably the assumed transmutation of the elements), and which hence concern the early history of chemistry. Having done with theory, we now have to inquire to what extent the ancients were acquainted with practical chemistry, what metals or other elements were known to them, and what processes depend upon chemical action. We do not, of course use the term "practical chemistry" strictly in its present use, because chemistry as a science was altogether unknown to the ancients. Some have, indeed endeavored to prove that the Egyptians must have been acquainted with the science, from the skill with which they used various metallic oxides for coloring the glass; but we have no proof of this.

Neither Herodotus nor Pliny, nor Vitruvius, indicates any knowledge of chemistry, as a science, among either Egiptians, Greeks, or Romans. Pliny, in his celebrated "Natural History," has laboriously amassed all the practical science and pseudo-science which the ancients possessed, and we find no mention of either chemistry or alchemy. At the same time it is impossible that the Egyptians and Sidonians can have attained their marvelous skill in the manufacture and coloring of glass, and in the extraction and working of metals, without the acquirement of a considerable amount of knowledge of the properties of matter, and of certain chemical changes. But this knowledge could never be worked up into a comprehensive system; it resulted from the labor of artisans, and the gulf between the philosopher and the manipulator was both wide and deep. There could be no union of practice and theory. Between Herakleitos with his theory that fire is the primal element, the actuating force of the Universe, and the man who wrought metals never so deftly, who applied fire to the use and service of mankind, there was no sympathy, no reciprocal transference of ideas.

To reason concerning the properties of matter with ones eyes shut was all very well, but to experiment with matter, to endeavor to determine the cause of such and such a change by experiment, was utterly unworthy of a philosopher. Anaxagoras is said to have made an experiment to prove that there is no vacuum. Aristotle found that a bladder of air weighed in air weighed more than the empty bladder (which if the experiment be properly made, is by no means the case), and hence concluded that the air has weight. But these are solitary exceptions; the way to study Nature, if she is to be studied at all, is, they maintained, to apply the pure, unaided intellect to the study, and to keep mind and matter as distinct as possible. From all this it resulted that your workers in metals and in curious arts, your makers of glass and pigments, kept their knowledge of matter to themselves, as secrets to be handed down from father to son.

Seven metals were known to the ancients, viz., gold, silver, copper, tin, iron, lead, and mercury. The first six are mentioned by Homer, and appear to have been known from remote antiquity, while mercury was not known till a later date; it was, however, common in the first century B.C. The Greek word μετάλλων, whence metallum and metal, signifies a mine, hence it was applied to anything found in mines, notably metals; μετάλλον metalon is connected with μετάλλαω - medal, "to search for diligently."

Gold has been valued from the earliest ages, on account of the peculiarity of its color, its luster, and its unalterably in air. The metal is invariably found in the native state, that is, uncombined with other substances, hence no metallurgical operation is necessary for its extraction. It is very often met with in surface deposits, and in early times was undoubtedly far more common in alluvium and the beds of rivers than now. It would thus be easily extracted by washing, and the grains could readily be fused together into a mass. Gold mines formerly existed in Ethiopia, in which the gold was found in a matrix of quartz, like much of the Australian gold of the present day. These mines were worked by the Egyptians, who employed large gangs of slaves for the purpose. The quartz was crushed, and the gold obtained from it by washing. We find representations of gold washing's, and the subsequent fusion of the metal, on Egyptian tombs, at least as early as 2500 B.C., that is to say, about the time of Joseph in Hebrew history. The woodcut (Fig. 2) is given by Sir Gardner Wilkinson, and is taken from a tomb at Beni Hassan. It represents gold washing, and the fusion and weighing of the metal.

Fig. 2.—Gold Washing: Fusion and Weighing of the Metal, from early Egyptian Tomb.

Gold Washing

Fig. 2.—Gold Washing: Fusion and Weighing of the Metal, from early Egyptian Tomb.

It is obvious that the process is only indicated, and not accurately or minutely portrayed. Another form of furnace is depicted in Fig. 3, and a blowpipe somewhat different from that shown in Fig. 2. The raised portion of the furnace is doubtless for the purpose of concentrating the heat upon the crucible, on the principle of the reverberating furnace.

Fig. 3.—Furnace and Blowpipe from Egyptian Tomb.

Gold once obtained was soon made into ornaments, very fine gold wire was used by the Egyptians for embroidery 3,300 years ago. Many of the Egyptian and Etruscan gold ornaments are very beautiful; we may notice particularly the gold myrtle wreath found in an Etruscan tomb a few years ago. The Egyptians also used gold for inlaying, and it was beaten into leaf and used for gilding as early as 2000 B.C. In the Odyssey the gilding of the horns of an ox about to be sacrificed is mentioned.

Silver like gold, is often found native, and from several of its ores, the metal may be extracted by the action of heat alone. It has been known from the earliest ages, and was used chiefly for ornaments and embroidery. Gold was used for money before silver, which was first known as "white gold." The oldest silver Greek coin is a coin of Ægina, and was, perhaps, coined in the eighth century B.C. But the oldest coins in existence are the electrum staters of Lydia. Electrum consists of about three parts of gold to one of silver. Probably the metals were first found in nature thus alloyed, and as no method of separating them was then known, they were worked up together. Electrum (ἤλεκτρο) was so called from its resemblance as regards color to amber (ἤλεκτρον ēlektron), which received its name from "beaming Sun"," ἠλέκτωρ (ēlektōr)", the sun. It will be remembered incidentally that the science of Electricity was so called by Gilbert of Colchester, because the attractive force was first observed in amber. Amber is mentioned more than once by Homer. Electrum as a metal is first mentioned in the Antigone of Sophocles. It was found naturally alloyed, as in the pale gold of the Pactolus, which contains a good deal of silver; and was also made artificially. Probably all very pale gold was called electrum; Pliny states that gold containing a fifth part of silver is called electrum. In the British Museum there are many coins made of this alloy.

Copper was in use before iron. It is, as is well known, usual to denote various early ages by the substances then used for domestic implements. Thus we have the "age of stone", the "age of iron", &c. The stone age is followed by the age of copper, this by the age of bronze, and the age of bronze by the age of iron. Homer wrote in the age of copper; the shield of Achilles is made of gold, silver, tin, and copper; the arms and implements and utensils of his heroes are of copper. Mr. Gladstone has argued at some length that by chalcos (Χαλκος) Homer meant copper, not bronze, as it is so often rendered. Chalcos is spoken of as a cheap and common metal, while tin was very scarce and rare; and it is scarcely probable could have contained ten or twelve percent of tin. Again, Mr. Gladstone points out that ipeaks of chalcos as ipvffpov, red, a term Id not apply to bronze; and he goes so say; "If chalcos be not copper, then copper mentioned in Homer" (Juventus Mundi, p. 530). At the same time we must remember that copper is very soft for cutting instruments, and a small quantity of tin hardens it. Some of the Greek bronzes only contain 1 per cent, of tin. Dr. Percy found in a bronze bowl of great antiquity from Nineveh, copper 99*51, tin 0*63. Ancient nails have been found containing copper 97*75, tin 2*25; and Mr. Gladstone suggests that, as tin is sometimes found associated with copper in nature, this may account for their composition. Copper is sometimes found native, sometimes in the form of ores, from which the metal is easily extracted. It appears to have been both cheap and plentiful at an early date. Romulus is said to have coined copper; it was also used for money by the Egyptians. Great confusion exists among old writers regarding the words signifying bronze and copper; Pliny clearly did not understand the difference between copper and bronze. The words χαλκός and χάλκινο; appear to have been applied indiscriminately both to copper and to alloys of copper containing a large proportion of that metal. Copper was alloyed with tin at an early date, because copper is soft and is unsuitable for cutting instruments, while the addition of tin hardens it. The fusing point of copper is between that of gold and silver, and is far below that of iron, while the fusing point of tin is only 446 deg. Fahr. Thus the two metals could be alloyed without any special metallurgical difficulties or the requirement of an inordinate temperature.

Copper was first obtained by the Romans from Cyprus, where it was very plentiful; they called it Æs Cyprium which became corrupted into Cuprum from which we get our present chemical symbol for copper, Cu. According to Solinus as was found at Chalkis, in Eubcea ; hence Χαλκος, the Greek word for copper. We read of "ores of æes," and of brass and bronze being dug out of mines, whereas the term brass is applied by us to an alloy composed of copper and zinc, and bronze to an alloy of copper and tin. Zinc as a metal was unknown to the ancients, and brass appears to have been made in Pliny's time by heating together metallic copper, calamine (a native carbonate of zinc), and charcoal; the latter reduces the calamine, and the metallic zinc and copper then combine. According to Dr. Thomas Thomson, aurichalcum or golden copper, was the proper name for brass. Ais is to be always translated copper or bronze, not brass, of which latter very little use appears to have been made. Among other alloys of copper, the ancients possessed the celebrated Als Corinthiacuniy which according to Pliny was formed accidentally during the burning of Corinth, by Mummius, B.C. 146. There were four varieties of this, one of which contained equal proportions of gold, silver, and copper; the others were most probably various admixtures of copper and tin. The commonest kind of ancient bronze contained in 100 parts, 88 parts of copper, and 12 parts of tin. Two specimens of bronze from Nineveh were found by. Dr. Percy to contain respectively:

	Bronze hook.	A small bell.
Copper	89·85	84·79
Tin	9·78	14·10
	99·63	98·89

Table 1

The proportion of copper and tin (about 10 to 1) is, remarks Mr. Layard, the composition of our best modern bronze, while the increase of tin in the case of the bell proves that the Assyrians were well acquainted with the increase of sonorousness produced by changing the proportions of the metals. Modern bell-metal contains about 80 parts of copper to 20 parts of tin. Sometimes a small quantity of lead was introduced by the ancients into their bronzes. Thus, a certain bronze for statues was formed by fusing together 100 parts of copper, 10 parts of lead, and 5 parts of tin. In a very ancient bronze armlet (probably Phoenician) found in this country, and belonging to a period anterior to the Roman occupation. Prof. Church found:

Copper	86·49
Tin	6·76
Zinc	1·44
Lead	4·41
Oxygen and loss	·90
	100·00

Table 2

Bronze was very much used in Egypt for vases, mirrors, arms, &c. These, according to Sir G. Wilkinson, usually contain from 80 to 85 per cent, of copper, with from 15 to 20 per cent, of tin. By the use of some acid substance, the surface was sometimes covered with a green or brown patina. Although the casting of the metals was not known in Greece in the time of Homer, bronze was probably cast in Egypt 2000 years B.C.

Several compounds of copper were used by the ancients: both the red and black oxide were obtained by heating copper to redness, and allowing it to cool in the air; they distinguished between the scales which fell off during cooling, and those which were caused to fall off afterwards by the blows of a hammer. These oxides were principally used for coloring glass. Verdigris or acetate of copper was obtained then, as now, by covering plates of copper with the refuse of grapes after the expression of the vine juice. Copper pyrites and a rude kind of sulfate of copper would appear from Pliny's obscure account to have been also known.

It follows from the above remarks concerning bronze, that tin, like copper, was known at a very early date. This is the more remarkable, because it has always been a comparatively scarce metal, and it was obtained from distant localities. Formerly it was almost entirely supplied by Spain and

Britain. The Phoenicians, who were the earliest traders, obtained it first from India and Spain, and afterwards from Britain. The Greek name for tin, kassiteros (κασσιτερίτη) was perhaps derived from the Insulæ...

> The word Kao-alrcpos is used both by Homer and Hesiod, and it is possible that it may have been borrowed from the Sanskrit kasttra, and, that tin was first procured from India. The Sanskrit word for tin, kastiray is clearly related to the word kds to shine. It is strange that the Arabic word for tin is kasdir, closely resembling the Sanskrit, although there is no family relationship between the languages. Possibly the Phoenicians first procured tin from India, and gave it a name resembling, its native name kastira; then the Greeks converted the Phoenician word into Kcuro-fofpost the Romans borrowed the word from the Greeks, and the fact of the scarce metal being found in certain islands north of Spain, was sufficient to secure for them the distinctive title of Insula cassiterides, or Tin Islands.

...Cassiterides, or Sicily Islands, from whence the Phoenicians asserted that they procured tin; but it has been suggested that in all probability they invented the story because they desired a monopoly of the metal, while in reality they procured all their tin from the mainland of Cornwall, where it has always abounded. Tin must have been very valuable or the Phoenicians would not have traded so far for it. Homer evidently considers it of far greater value than copper. In the time of Pliny it was worth about eight shillings the pound. The metal was t known in Egypt 2000 B.C. Pliny mentions that it was found in the form of small black grains in alluvial soils, from which it was obtained by washing; this account would agree with a description of the so-called stream tin, which is tin ore separated from the parent vein, and carried down by streams. It is an oxide of tin, and the metal is obtained from it by strong ignition with charcoal. Tin was used for tinning copper vessels, for making mirrors, and in the manufacture of bronze. In the Iliad the graves of the armor of Achilles are made of tin, and it enters into the composition of the shield; it was also used for coating copper.

Chapter IV.

Iron - Lead - Quicksilver - Colors used for Painting and Dyeing - Glass - Certain Minerals known to the Ancients - Miscellaneous Processes,

> Iron was not in common use till long after the introduction of copper. It is far more difficult to procure, because it is not met with in the native state, and the fusing point is very high. The metallurgy of iron is more complex than that of copper, and when obtained it is a more difficult metal to work. According to Xenophon the melting of iron ore was first practiced by the Chalubes, a nation dwelling near the Black Sea, hence the name Chalups (Χάλυβα) used for steel, and hence our word Chalybeate applied to a mineral water containing iron.

Steel was known to the ancients, but we do not know by what means it was prepared; it was tempered by heating to redness, and plunging in cold water. According to some, kuanos (κύανος) mentioned by Homer was steel (rather blue, cyan); but Mr. Gladstone prefers to conclude that it was bronze. Iron was known at least 1537 B.C. It was coined into money by the Lacedaemonians, and in the time of Lukourgos was in common use. It was used in the time of Homer for certain cutting instruments, such as woodman's axes, and for plough shares. Its value is shown by the fact that Achilles proposed a ball of iron as a prize for the games in I honor of Patroklos. Neither iron money nor iron implements of great antiquity have been found, because, unlike the other metals of which we have spoken above, iron rusts rapidly, and comparatively soon disappears. No remains of it have been found in Egypt, yet Herodotus tells us that iron instruments were used in building the pyramids ; moreover, steel must have been employed to engrave the granite and other hard rocks, massive pillars of which are often found engraved most delicately from top to bottom with hieroglyphics. Again, the beautifully engraved.

FIG. 4.—Egyptian Bellows. Fifteenth Century B.C.

Bellows

Fig. 4.- Egiptians Bellows. Fifteenth Century B.C.

Babylonian cylinders and Egyptian gems, frequently of carnelian and onyx, must have required steel tools of the finest temper. We have no record of the furnaces in which iron ore was smelted, but we know that bellows were in use in the 15th century B.C. in Egypt, and some crucibles of the same period are preserved in the Berlin Museum. They closely resemble the crucibles in use in the present day. The accompanying woodcut (Fig 4) represents a double pair of bellows, a furnace, fuel, and perhaps a crucible.

The native Indians prepare iron from hematite at the present time by equally primitive bellows, which indeed resemble the above very closely, and which, without doubt, have been unaltered for centuries. A small furnace, A (see the accompanying section, Fig. 5), is rapidly constructed of clay, and into the bottom of this two nozzles, are introduced at B; these are connected with the bellows by bamboo tubes. The bellows, c, consists of a cup-shaped bowl of wood covered with goat-skin above, and connected with the bamboo below. In the center of the goat skin cover a round hole is cut; the blower places his heel upon this, which is thus closed, while at the same time the skin is depressed and a blast is driven from the tube, then he steps upon the second skin, and thus a nearly continuous blast is kept up. The bent bamboo and string, D, is for the purpose of raising the goat-skin cover of the bellows after depression, which, it will be noticed, is accomplished in the Egyptian bellows by a string raised by the hand. A piece of hematite is introduced with some charcoal, and after the lapse of some time, it is reduced by the carbonic oxide to a spongy mass of iron. Undoubtedly a crude furnace and appliance of this nature was used by the first smelters of iron.

FIG. 5 —Smelting Furnace and Bellows used by native Indians in the present day.

Furnace

Fig. 5 - Smelting Furnace and Bellows used by native Indians in the present day. We are indebted to Dr. Percy for permission to copy this figure from his "Metallurgy," and to Mr. Murray for Figs. 2, 3, 4, and 5.

Although we hear less of lead than of the preceding metals, it was known to the Egyptians at an early date, and it is mentioned by Homer. In the time of Pliny leaden pipes were used to convey water; and sheet lead was employed for roofing purposes. The chief supply of the metal came from Spain and Britain. Pliny believed that lead was reproduced in the mine, so that if an exhausted mine were closed it would be fit to work again in a few years' time. This idea of the growth of the metals was very generally accepted by the alchemists. Tin and lead were sometimes alloyed together by the ancients, and tin was used as a solder for lead. Litharge or protoxide of lead, and cerussa usta (burnt ceruss), or red lead, were used by painters. Cerussa, which we now call "white lead," or, more strictly, carbonate of lead, was prepared by exposing sheets of lead to the fumes of vinegar in a warm place, a heap of decomposing manure, for instance. A basic acetate of lead is formed by this means, which is partially converted into carbonate by the carbonic acid given off by the decomposing organic matter. Cerussa was used by Athenian ladies as a cosmetic Cerussa testa was first formed accidentally from cerussa during the burning of a house near the Piraeus. Litharge is easily formed by heating lead above its melting point in air, when it absorbs oxygen gas, and the resulting oxide may be skimmed off.

Mercury was common in the time of Pliny, but it is not mentioned by earlier writers. It was found native in Spain, but was more generally obtained by heating cinnabar (sulfide of mercury) with iron filings, in an earthen vessel, to the top of which a cover was luted. The iron decomposed the sulfide, and the liberated mercury was volatilized and condensed on the cover of the vessel, whence it was collected. This method, described by Dioscorides, is the first crude example of distillation, which afterwards became a principal operation among the alchemists and chemists for separating the volatile from the fixed. In the time of Dioscorides cinnabar was called minium, but it became so largely adulterated with red lead that the term minium was ultimately applied to the latter. Minium is still one of the names for red lead. Pliny was acquainted with the high specific gravity of mercury, and with its power of dissolving gold. Substances were sometimes gilded by a gold amalgam. Mercury was also used, as now, for extracting gold from its earthy matrix; the gold-bearing rock was powdered and shaken up with mercury, which dissolved out the gold; the amalgam of gold and mercury was then squeezed through leather, which separated most of the mercury; the solid amalgam was heated to expel the mercury, and pure gold remained. Vitruvius states that gold was recovered from gold embroidery by burning the cloth in an earthen pot, and throwing the ashes into water to which quicksilver was added. The latter attracted the gold and dissolved it; the amalgam was put into a piece of cloth and squeezed between the hands, and the mercury, on account of its fluidity, was forced through the pores of the cloth, while the gold remained.

Native mercury was called argentum vivum (quicksilver), while mercury distilled from cinnabar was called hydrargyrum (υδράργυρο, υγρó ασήμι, liquid silver), from which we take our present symbol for the metal, Hg. The alchemists, in whose eyes, as we shall hereafter see, mercury was a very important metal, call it by the various names of mercurius, argentum vivum, hydrargyrum, with others of a more fanciful nature.

The ancients were not acquainted with any other metals in an uncombined state, except the seven mentioned above. Stibium, or sulfide of antimony, was used in the East at an early period for painting the eyelashes. It is still used for that purpose, and is called kohl. Native carbonate of zinc was known, and black oxide of manganese. The two sulfides of arsenic were known, and were used as pigments. The yellow sulfide was called auripigmentum and arsenicum; the red sulfide went by the name of sandaracha. Auripigmentum became contracted into orpiment, a word which we find both in alchemical treatises and in our most modern treatises on chemistry.

The colors used by the ancients for painting were, examined by Sir Humphry Davy at the beginning of this century, and he came to the conclusion that "the Greek and Roman painters had almost all the same colors as those employed by the great Italian masters at the period of the revival of arts in Italy." Various colors have been examined from the frescoes in the Baths of Titus, from Pompeii, and from Egyptian tombs. The colors of the Egyptians were red, yellow, blue, green, black, and white. The red was bole, that is a clay deriving its color from oxide of iron; the yellow, an ochre, also clay, colored by a paler form of oxide of iron; the green, a mixture of this ochre with a blue powdered glass, produced by fusing together sand, carbonate of soda, and oxide of copper. The black was ivory black, prepared by heating bones out of contact with air until completely carbonized ; the white was powdered chalk. These various colors were mixed with gum and water before use. The Greeks and Romans used red lead and cinnabar, as well as red ochre, and yellow protoxide of lead. The blue powdered glass mentioned above was sometimes called κυανός by the Greeks, Cæruleum by the Romans. Vitruvius describes the method of preparing it; and Davy prepared a substance which perfectly resembled the ancient color, by fusing together fifteen parts of carbonate of soda, with twenty parts of powdered flints, and three parts of copper filings.

The green of the Romans was carbonate of copper, and for browns they sometimes used dark oxide of manganese. The purpurissum of the Romans was Tyrian purple, a very valuable color obtained from a shell-fish, and much used for dyeing. In order to obtain the color for the purposes of painting, clay was placed in the cauldrons of dye, so as to absorb the color, and was afterwards removed and dried. Indicum purpurissum was probably indigo; Pliny mentions that the vapor possesses a fine purple color. Ivory black was called Elephantinum; lamp black, that is soot, was called Atramentum. The latter mixed with water constituted the ink of the ancients.

According to Pliny, glass was first discovered by some Phoenician merchants who were returning from Egypt with a cargo of natron (carbonate of soda), and who landed on the sandy banks of the river Belus. In order to support the vessels they used for cooking their food over the fire, they used some large lumps of natron, and the fire was sufficiently strong to fuse it, with the fine sand of the river. Hence resulted the first glass. Whatever may be the value of this story, we find representations of glass-blowing on the monuments of Thebes and Beni Hassan; and the Egyptians were well acquainted with it 2450 B.C. The most celebrated manufacture of glass was in Egypt; and, according to Strabo, a peculiar kind of earth found near Alexandria was essential for the finer kinds of glass. The Egyptian glass had nearly the same composition as our "crown glass," which contains 63 per cent, of silica, 22 of potash, 12 of lime, and 3 of aluminia, The Phoenicians and Egyptians exported large quantities of glass to Greece and Rome. The Egyptians engraved and cut glass with the diamond; they also possessed extraordinary skill in coloring glass with various metallic oxides, and combining several colors in the same vase, and they imitated precious stones with great success. We read of whole statues made of emerald, but these were undoubtedly of emerald glass, viz. glass colored by oxide of copper. The Egyptians understood the art of enameling on metal. Aristophanes is the first Greek author who mentions glass (rrfv vdkov); he alludes to the use of a lens of glass, as a burning-glass in the NeXai, which play was acted in Athens, B.C. 423. colorless glass was the most valuable, and a small quantity of oxide of manganese was added then as now for the purpose of decolorising it. A very ancient opaque green glass, analyzed by Klaproth, was found to contain 65 per cent, of silica, 10 of oxide of copper, 75 of oxide of lead, 3*5 of oxide of iron, and about 6 per cent, of both lime and alumina. A red glass was found to be colored by red oxide of copper.

Dyeing was much practiced by the ancients; the Egyptians understood the effect of acid on some colors, and were acquainted with mordants, that is, substances which "fix" the coloring matter in the fabric, and prevent it from being washed out. The . most celebrated dye of antiquity was the purple of Tyre, discovered about 1500 B.C., perhaps earlier. It was produced by certain shell-fish which inhabit the Mediterranean; these are spoken of as buccinum and purpura by Pliny. A few drops only of the dye were obtained from each fish, and the color hence became very valuable, and was monopolized by the emperors of the world. The Egyptians dyed linen with indigo, which they procured from India, for they had considerable intercourse with that country at an early period.

Lime was used for removing the hair from skins about to be tanned. Leather made in the time of Sheshonk, the contemporary of Solomon, has been found in a good state of preservation. For the process of tanning, they used the pods of the Acacia Nilotica, a plant which, according to Sir G. Wilkinson, was also prized for its timber, charcoal, and gum.

Nitrum was a term applied to carbonate of soda, or natron, which, we have already seen, was used in the manufacture of glass. The substance which we now call nitre (nitrate of potash) was probably known in India and China before the Christian era. Dr. Thomas Thomson has suggested that when the real nitre was imported into Europe, it received the same name as carbonate of soda (nitrum), from the similarity of its appearance, and retained the name on account of its greater importance. Roger Bacon always speaks of nitrate of potash as nitre. The low Latin name for soda became natrium, hence our present symbol for sodium, Na,

Soap is first mentioned by Pliny ; it was made by mixing wood ashes, which contain carbonate of soda, with animal fat. It was used solely as a kind of pomatum. The Greeks added wood ashes to water to increase its cleansing properties.

The only acid with which the ancients were acquainted was acetic acid or vinegar. It has been suggested that the Egyptians discovered nitric acid and nitrate of silver, because a silver stain has been found upon some linen, but the evidence is insufficient. We remember the story of Cleopatra dissolving two pearls, valued at ten millions of sestertii, in vinegar; although only a story, it would seem to show that vinegar was the most powerful solvent known. This is further indicated by the story of Hannibal dissolving rocks by vinegar.

A number of minerals are mentioned by Pliny, but we can recognize but few of them. Iron pyrites (sulfide of iron) was used for striking fire with steel in order to kindle tinder, and was hence called pyrites (irvp, fire), or fire-stone. Sulfur was well known, and was used for matches; it was also apparently burnt in a current of air, and the sulfurous acid produced employed was for bleaching purposes. Asphalt was used for embalming, and undoubtedly also for torches.

Chapter V.

Association of the seven Metals with the seven greater Heavenly Bodies - Consequent introduction of Symbols into the history of Matter.

Thus far we have become acquainted with the various theories of the Ancients, in which changes in the composition of matter are discussed, and with various processes by which changes were actually effected. Before we leave this period, and pass at one bound to the eighth century A.D., we must notice the commencement of a symbolically system in the history of matter, which in the hands of the Alchemists and early Chemists assumed vast proportions, and still appertains to the science of Chemistry. This system was commenced by the association of the seven metals with the seven greater heavenly bodies. We do not know at what period the metals were designated by the names and symbols of the planets: certainly in a very remote age.

At a very early date the Chaldaeans represented the stars by symbols, and these gradually increased until astrology became one mass of symbols. On the occasion of certain religious ceremonies the Kings of Assyria wore a necklace in which the sun, moon, and stars were represented as emblems, for they were first worshiped as emblems of the Deity. Sculptural representations of necklaces with seven discs upon them have also been found. Symbols were carried before Egyptian priests, and their gods were, represented with certain signs symbolical of their special attributes.

The Assyrian goddess Astarte carries in her left hand a symbol (b) (Fig 6) not very different from the crux ansata of the Egyptians (a); and the symbol (c) by which the planet Venus was afterwards represented by the astrologers, and is still represented by astronomers. In the celebrated "Book of the Dead" (B.C. 1350), the most perfectly preserved Egyptian ritual which the world possesses, this latter symbol (c in the figure) occurs frequently among the hieroglyphics. This is very noticeable in the "Judgment scene" of the Turin papyrus, a copy of which exists in the British Museum. The upper portion of the crux ansata was frequently made more rounded in form, and it is obvious that if, in addition to this, the cross was somewhat lowered, we should arrive at the third symbol (c) shown above. The crux ansata (a), if written quickly, could easily pass into this latter symbol (c), and this may account for the occurrence of both symbols in the judgment picture, to which we alluded above.

Fig. 6.— a Crux ansata of the Egyptians; b Assyrian symbol of Astarte; c Later symbol of the planet Venus.

Symbol

Fig. 6.— a Crux Ansata of the Egyptians; b Assyrian symbol of Astarte; c Later symbol of the planet Venus.

Plato speaks of the sun, moon, and five planets, but does not distinguish them by the names of gods; Epinomis mentions them in conjunction with the names of gods. It is probable that the Chaldaeans also associated the principal heavenly bodies with the names of deities — San with the sun, Hurki with the moon, Bel-Merodach with Jupiter, Astarte or Ishtar with Venus, Nergal with Mars, &c. The relative position of the planets was generally as follows: the Earth was the center of the system; next in order came the Moon, the Sun, Venus, Mercury, Mars, Jupiter, and Saturn; but these positions were sometimes varied. It was known that Saturn completed a revolution in about thirty years, while Jupiter required twelve years, Mars only two, and Mercury and Venus appeared to take about the same time as the Sun; hence the above order. As Saturn was farthest from the source of heat, and the slowest in his motion, he was supposed to be of an icy character, and to assert an evil influence.

While speaking of the seven greater heavenly bodies, and the seven metals, we may allude incidentally to the curious prominence of that number in many matters "that mysterious number," as Mr. Layard calls it, "so prevalent in the Sabaean (or Sabaic) system." Thus (to select a few instances at random) we have seven days of the week, seven wise men of Greece, seven wonders of the world, seven cardinal sins, seven-stringed lyre, seven harmonic proportions, seven heavens, seven walls of Ecbatana, seven gates of Thebes. The list might be extended almost indefinitely. Among the Hebrews the number was specially prominent. Not to mention the frequent allusion to it in the Apocalypse, we may recall the incidents of the fall of Jericho: the town was surrounded for seven days; on the seventh day the walls fell at the blast of seven trumpets, which were carried round the walls seven times by seven priests.

We cannot tell why the seven metals were associated with the seven deified heavenly bodies, unless it was because all things which amounted to the same number were connected with them. This, at least, we know, that long before the time of Geber, the first writer on chemistry, the metals had received the same names and symbols as the planets. "There is abundant evidence," says Mr. Gladstone, "of a correspondence between the seven metals of Homer and the seven metals of the ancient planetary worship of the East." In the time of Homer only six simple metals were known, and the seventh was the compound kuanos (kuanos ("blue ore used in dyeing"), denoting by metonymy the dark blue color. Attested since Homer (probably borrowed from Hittite); quicksilver afterwards became the seventh simple metal, and received the name and symbol of the seventh planet The metals were apportioned as follows:

Gold	The Sun	☉
Silver	The Moon	☽
Quicksilver	Mercury	☿
Copper	Venus	♀
Tin	Jupiter	♃
Iron	Mars	♂
Lead	Saturn	♄

Seven

Herodotus tells us that Ecbatana had seven walls, the outermost of which was the lowest, and the others gradually ascended like steps to the highest, which enclosed the king's palace. They were each painted of a particular color; the outermost. white, the second black, the third purple, the fourth blue, the fifth red, the sixth the color of silver, the seventh the color of gold. Undoubtedly these had reference to the seven greater heavenly bodies. It is impossible to account for the Colors, but it is curious to notice the particular color which would fall to any particular metal. Placing the planets in order as applied to the metals, we should have gold to gold, silver to silver, red to copper, blue to iron, purple to tin, black to lead, the most despised of the metals. It is probable that the Sabaeans associated these colors with the seven heavenly bodies. The temple of Bel-Merodach, rebuilt by Nebuchadnezzar, and called by him the " Wonder of Borsippa," appears also to have consisted of seven terraces differently colored. The following is a portion of the inscription from a clay cylinder found among the ruins of the temple:—" I (Nebuchadnezzar) have completed the magnificence of the tower with silver, gold, precious stones, enameled bricks, fir, and pine: .... This most ancient monument of Borsippa is the house of the seven lights of the earth."

How the symbols conferred upon the planets and afterwards upon the metals arose it is difficult to say; they are undoubtedly of Chaldean origin, but to what extent they have since been modified no one can tell. They exist in early MSS. on Alchemy.

That the sun should be represented by a circle, the

• symbol of perfection, is no wonder. Again, that the moon should be symbolized by a crescent we can understand ; but the others present greater difficulties.

• Among these, some say we have the looking-glass of Venus, the thunderbolts of Jupiter, the spear and shield of Mars, the scythe of Saturn, and the caduceus of Mercury. In the temple of Hermes at Pselcis, he is represented with a staff having a serpent twining around it, from which it has been suggested the caduceus of Mercury may have been derived. (See Fig. 7, p. 65.) Some see in IT, not the thunderbolts, but the throne of Jupiter; others the Zeta of Zeus; others, again, the Arabic 4, indicating that Jupiter was the fourth planet in order. Some, too, have seen in h the K of Kronos. It is less difficult to understand why a particular metal was assigned to a particular heavenly body. Thus gold would naturally be associated with the sun, on account of its color, perfection, and beauty, and because it was ever regarded as the noblest metal. For the same reason silver would fall to the moon, with its pale, silvery color and light. So, again, iron, the metal of war, would be associated with Mars; lead, the dull, despised metal, with Saturn, the slowest of the planets; quicksilver, the nimble volatile metal, with Mercury, the messenger of the gods.

These signs became in the hands of the Alchemists the commencement of a symbolic system in chemistry.

Chapter VI.

The Alchemists — Origin of Alchemy — Hermes Trismegistus — Greek MSS. on Alchemy — Their probable authorship and age.

We spoke in the last chapter of the alchemists almost for the first time, and we must now turn our attention to the origin and growth of their dogmas, and to their work. We have already seen that the word χημεία is first found in the Lexicon of Suidas, and that he defines it as "the preparation of gold and silver." He further tells us, under the same heading, that the books on the subject, were sought for by Diocletian and burnt, lest the Egyptians should become rich through their knowledge of the art, and should thus be able to resist the Romans. Now, the people who professed a knowledge of the art of making gold were called alchemists. The word alchemy, as we have previously shown, consists of a Coptic root united with an Arabic prefix, and signifies the hidden or obscure art.

Alchemists were those who practiced this mysterious art. We can well understand why the professors of such an art should maintain the utmost secrecy; to divulge such magic would be to make all men equally rich; hence it was necessarily a hidden art. Neither did the books on the subject avail much, for they are filled with some of the most incomprehensible nonsense that ever was written. Yet the literature of the subject is enormous. The volumes on alchemy in our large libraries are to be counted by the hundred. In 1602 Zetzner published, in Strasburg, a "Theatrum Chemicum," containing more than a hundred tracts on alchemy, selected from various notable authors. A century later Mangetus published his "Bibliotheca Chemica Curiosa," in two large folios, containing a hundred and twenty two alchemical treatises. We have previously given the titles of a few Greek MSS. on alchemy. The list has been extended to eighty-three. Arabic and Persian MSS. on the subject are not uncommon. There are treatises in Spanish, Italian, German, Dutch, and English on alchemy, and, more numerous than all, treatises in Latin, in every large library. Let us endeavor to get from the tangled mazes of this hieroglyphic literature some idea of alchemy, and of its influence upon chemistry.

We are, perhaps, puzzled at the outset to comprehend how any one man, much less thousands of men, could have deluded themselves with the belief in the possibility of transmuting one kind of matter into another: crude lead, or tin, or mercury, into weighty, lustrous gold. But this was not the greatest wonder of the age. At the time when alchemy arose, and throughout the period during which it most flourished, the belief in theurgy, witchcraft, necromancy, and magic of all kinds was rife among all classes; and surely it was less wonderful to change lead or tin into gold, than to call up the spirit of one's ancestor, or to confer perpetual youth upon a nonagenarian! It is, for wonderment, as compared with the greater magic of the day, as the process for the conversion of benzine into aniline compared with spirit-rapping; or as a demonstration of specific inductive capacity compared with a manifestation of psychic force. Alchemy was considered to be perfectly rational not two centuries ago, and was among the lesser forms of magic, inasmuch as it did not require the influence of supernatural causes.

The growth of the idea is not difficult to trace. The ancients had persistently asserted the change of one element into another. Thales, as we have seen, evolved the ten thousand forms of nature and kinds of matter, from water, Anaximenes from air, by successive transmutation. Aristotle, whose physical views were accepted without question by the alchemists, had endeavored to show by clever argument that, if you transfer a quality of water to fire, you obtain air; while if you transfer a quality of earth to air, you get water; and so for fire and earth, and that from these elements all things proceed. This was readily accepted by Middle Age thinkers. The alchemists reasoned, plausibly enough:—if fire becomes air, air water, and water earth, why may not one kind of substance formed from these elements be changed into another kind of substance of somewhat the same nature, and certainly more similar than air and water, or water and earth? Why may not lead, compounded of these elements in certain proportions, be changed into gold, compounded of these elements in certain other proportions? There have been falser modes of reasoning than this in the history of science.

Let the ancient Greek theory of the transmutation of the elements be once literally accepted, and the alchemical belief in transmutation follows naturally; it is a minor application of the major proposition. There is nothing to wonder at in this; the human mind seldom moves by fits and starts; an essentially new mode of thought and new form of belief is rare, and many apparently new dogmas are united with older dogmas in the closest manner, and are in fact direct emanations from them. Such was the alchemical idea of transmutation. Admitting the possibility of the process, a man would naturally ask himself "What do I most desire to make?". "What in this world procures the greatest amount of happiness and of power?" For what have men slaughtered each other by the thousand in open war, or singly and, secretly in the dead of night? For what have kingdoms been sold, great tracts of land ceded, and people been ground into serfdom till they rose and rioted against their oppressors? For what have princes and cardinals been created, emperors and kings destroyed, and the eternal peace of troubled souls promised? In a word, for what will man dare all things, sacrifice all things; for what will he toil during a lifetime; to what will he devote all his intellectual energies? This is surely the thing for the ready acquirement of which we may devote much time and thought, and this thing is gold. This is the key to the prodigious masses of alchemical literature, and to the mysteries and anomalies, connected with men who often wasted their whole lives and ail they possessed in the endeavor to change baser metals into gold.

If we consult alchemical MSS., no matter the date or author, or language, we find constant mention of Hermes Trismegistus, who was indeed considered, and sometimes designated, the father of alchemy. In a treatise attributed to Albertus Magnus we are told that the tomb of Hermes was discovered by Alexander the Great, in a cave near Hebron. In this was found a slab of emerald which had been taken from the hands of the dead Hermes by Sarah, the wife of Abraham, and which had inscribed upon it in Phoenician characters the precepts of the great master concerning the art of making gold. The inscription consisted of thirteen sentences, and is to be found in numerous alchemical works. It is for the most part quite unintelligible, and in style closely resembles the great mass of Middle Age alchemical literature.

The following is cited as the inscription of the "Smaragdine Table," and is to be found in very early MSS. in various languages:

1. I speak not fictitious things, but that which is certain and most true.

2. What is below is like that which is above, and what is above is like that which is below, to accomplish the miracles of one thing.
3. And as all things were produced by the one word of one Being, so all things were produced from this one thing by adaptation.
4. Its father is the sun, its mother the moon; the wind carries it in its belly, its nurse is the earth.
5. It is the father of all perfection throughout the world.
6. The power is vigorous if it be changed into earth.
7. Separate the earth from the fire, the subtle from the gross, acting prudently and with judgment.
8. Ascend with the greatest sagacity from the earth to heaven, and then again descend to the earth, and unite together the powers of things superior and things inferior. Thus you will obtain the glory of the whole world, and obscurity will fly far away from you.
9. This has more fortitude than fortitude itself; because it conquers every subtle thing and can penetrate every solid.
10. Thus was the world formed.
11. Hence proceed wonders, which are here established.
12. Therefore I am called Hermes Trismegistus, having three parts of the philosophy of the whole world.
13. That which I had to say concerning the operation of the sun is completed.

The story and the inscription, together with all books attributed to Hermes (who is asserted to have lived about 2,000 B.C.), are no doubt the production of monks of the Middle Ages. In spite of the obvious worthlessness of the inscription of the emerald table, men have not been wanting who have labored long and lovingly to prove its authenticity, to interpret it, and to show that it is in good sooth a marvelous revelation, full of sublime secrets of considerable import to mankind.

Hermes Trismegistus is generally asserted by the alchemists to have been a priest who lived a little after the time of Moses. According to Clemens Alexandrinus he was the author of forty-two books containing all the learning of the Egyptians; others tell us that he was the author of several thousand volumes. Plato speaks of him in the "Phædrus" as the inventor of numbers and letters. He was in fact the Egyptian god of letters, and as such of course could be described as the author of multitudinous works. He was the deified intellect, and hence has often been confounded with Thoth, "the intellect" Sir Gardner Wilkinson speaks of Hermes as an emanation of Thoth, and as representing "the abstract quality of the understanding." The woodcut (Fig. 7) representing Hermes, is from a temple at Pselcis,- which was erected by Erganum, a contemporary of Ptolemy Philadelphus. It may be well to note the extent of the symbolism associated with the sculpture; in one hand Hermes holds the Crux ansata, the symbol of life, in the other a staff associated with which are a serpent, a scorpion, a hawk's head, and, above all, a circle surrounded by an asp, each with its special symbolic significance. On the Rosetta stone Hermes is called "the great and great", or twice great; he was called Trismegistus, or thrice great, according to the twelfth aphorism of the Emerald Table, because he possessed three parts of the wisdom of the whole world, which in his light of deified intellect he might well do.

Fig. 7 - Hermes Trismegistus: from the temple at Pselcis

Perhaps no author is more often quoted by the alchemists than Hermes, the supposed father of their art. They called themselves Hermetic philosophers. Alchemy is often called the Hermetic Art, or simply Hermetics. To enclose a substance very securely, as by placing it in a glass tube and fusing, or sealing, the mouth of the tube, was called securing with "Hermes his seal," and the echo of the idea lives among us yet; for, in our most modern treatises, the expression "to seal hermetically" may be found. Petrus Hauboldus, of Copenhagen, was surely one of the most enterprising publishers of his day, for he had the temerity to publish a book entitled Hermetis Ægyptiorum et Chemicorum Sapientia. A book square as to its dimensions, small as to its type, drier than dust as to its contents, of four hundred odd pages, of two centuries of age, writ in Latin, with a sprinkling of contracted Greek, and floridly dedicated to Jean Baptiste Colbert. A book wherein the author endeavor to prove that alchemy was known before the flood, that Hermes Trismegistus was a real personage, the inventor of all arts, the father of alchemy, and much else besides. We may well imagine that the author of such a treatise was no ordinary man, and our conjecture proves a tolerably correct one.

Olaf Borch, whose Latinised name became the more resounding Olaus Borrichius, was apparently the great mainstay of the University of Copenhagen; at all events, he was simultaneously Professor of Philology, Poetry, Chemistry, and Botany; and we must either imagine that, in 1660, professors were difficult to procure in the Kingdom of Denmark, or else that Olaus Borrichius was such an astounding genius that he could readily undertake the duties of four diverse professorships at the same time.

We can scarcely imagine three greater antitheses than the philological faculty, the poetical faculty and the chemical faculty; but here we find them united, or assumed to be united, in one man. Yet more, Borrichius was appointed Court Physician, and Assessor of the Supreme Court of Law. He was the very personification of all learning, if we may judge by the treatment he received from his countrymen. In addition to the work mentioned above, he wrote various treatises on philology, on the quantity of syllables, on the Greek and Latin poets, on medicine, chemistry, and botany. It is strange that a man who, presumably in his capacity of judge, was in the habit of sifting evidence, and of avoiding hasty generalization, should have endeavored with much elaborate argument to prove that Hermes Trismegistus was a real personage; that his Smaragdine table was really found by the wife of Abraham, and that it contained matter of the highest import to mankind. We must imagine that in this matter Borrichius allowed the imaginative faculty due to his poetical temperament to exert an undue influence over his more sober judgment. He is equally at pains to assert the authenticity and antiquity of the various Greek MSS. on alchemy in the libraries of Europe. He specially mentions a MS. by Zozimus of Panapolis, on the art of making gold, in the King's Library in Paris; and Scaliger tells us that this same MS. was written in the fifth century. M. Ferdinand Hoefer is apparently penetrated by the Borrichian spirit of faith and imagination, and he unhesitatingly accepts the early date attributed to the Paris MS.

M. Hoefer traces the rise of Alchemy to the fourth century of our era; it was then known as the "sacred art" (ars sacra; Ιερά Τέχνη), and one of the chief writers on the subject was the said Zozimus of Panapolis. The principal Greek MSS. attributed to Zozimus, which exist in the Bibliotheque Nationale, have the following titles : (alpha) On Furnaces and Chemical Instruments ; (beta) On the Virtue and Composition of Waters ; (gamma) On the Holy Water; (delta) On the Sacred Art of making Gold and Silver. In the latter, Zozimus mentions that if the "soul of copper," which remains above the water of mercury, be heated, it gives off an aeriform body (

Democritus (not to be confounded with the Greek philosopher of that name), in his "Physics and Mystics," informs us how he invoked the shade of his master, Ostane the Mede, and how the spirit appeared and accorded him mystical communing. Synesius, the commentator of Democritus, lived, according to M. Hoefer, about fifty years after Zozimus (say 450 A.D.); but a treatise on the Philosopher's Stone is in existence which claims Synesius as its author, which mentions Geber, who lived at least 400 years later. Mary the Jewess, who is often alluded to by later alchemists, was a contemporary of Democritus, and a writer on alchemy; she also invented various chemical vessels, among others a bath, to gently transmit heat by means of hot sand or cinders, which (according to M. Hoefer) is still called, after her, a Bain-Marie.

We cannot assign to the Greek MSS. in the Bibliothèque Nationale the antiquity which M. Hoefer and others so readily accept; and we must still hold to our opinion that they and all other known Greek MSS. on alchemy are the production of later centuries, and are probably the work of Greek monks. In the first place, who was Zozimus ? Was it Zozimus the Anti-pope, who succeeded Innocent I., or Zozimus the Sophist of Alexandria, or Zozimus the historian? No one can tell. It cannot be pretended that any of the Paris MSS. are in the actual writing of Zozimus. One of them is entitled "Zozimus the Panapolite, on the Chemical Art, to his sister Theosebia;" but, according to the "Biographie Universale," it was Zozimus of Alexandria who dedicated books to his sister Theosebia, and he lived in the third century B.C., while Zozimus of Panapolis lived in the fourth century A.D. Here, then, we have a discrepancy of 700 years, and a clear confounding of Zozimus of Alexandria with his namesake of Panapolis. Suidas attributes chemical works to the former, but we must remember that the word χημεία does not occur before the eleventh century, A.D. The director of the Bibliothèque Nationale, in a recent letter for which we

This Library has so often changed its name of late, that we think it necessary to mention that we mean the Library in the Rue Richelieu, which is called by old writers the Bibliotheque du Roi sometimes the Bibliotheque Royale, lately the Bibliotheque Imperiale, still more lately the Bibliotheque Communale now the Bibliotheque Nationale. Juncker, in his Conspectus Chemia, in speaking of various writers on alchemy, cites "Zozinms Panapolites celeberrimus et magni cognomen adeptus, cujus varia scripta exstant in Bibliotheca Regia Parisiensi."

I have to thank him, writes as follows: "La Bibliothèque Nationale ne renferme aucun manuscrit grec de Zosime de Panapolis qui puisse attribu a une èpoque anterieure au XIII Siècle. Le plus ancien de ceux qu'elle possède ne remonte pas plus loin que cette date." Everything tends to prove that the MSS. were not only written, but composed, at a period posterior to the fifth century. The fanciful titles of some of them show us that their authors adopted any name they pleased; thus we have "the Epistle of Isis, queen of Egypt, and wife of Osiris, on the Sacred Art, addressed to her son Horus," in which we. find a solemn oath dictated to Isis by the angel Amnael, who swears by Mercury and Anubis, by Tartarus, the Furies, and Cerberus, and by the dragon Kerkouroboros. The whole thing is plainly a blending of Eastern and Western thought: personages of Egyptian, Greek, and Roman mythology, with angels of the Talmud and genii of Arabic lore. We are glad to find that M. Hoefer breaks freely away from the too confident Olaus Borrichius, as to the authenticity of Hermes Trismegistus. He admits that the books which bear his name are spurious, and concludes that their author "vivait probablement a l'epoque critique du Christianisme triomphant et du paganisme à l'agonie." But if we take this as the time of Constantine the Great, we must venture to attach a later date to these writings.

We recently had an opportunity of examining the MS. in the Bibliothàque Nationale, attributed to Zozimus and to the fifth century; a MS. which, from its frequent mention in both ancient and modern works on the history of chemistry, possesses special interest. It is entitled "Zozimus on Chemical Instruments and Furnaces, and on the Holy Water" (Zozimus για τα Χημικά μέσα και Φούρνοι, και στο Αγίασμα), and it is a well preserved MS. of the thirteenth century, written on vellum. The few drawings which it contains are asserted to have been taken by the author from a temple at Memphis.

Fig. 8.—An Alembic, and Symbols from Greek MSS. on Alchemy.

Fig. 8.—An Alembic, and Symbols from Greek MSS. on Alchemy.

The Alembic (b in the accompanying woodcut, Fig. 8) is copied from this MS., in which also the line of symbols (a) is found. These symbols occurred in almost every Greek MS. on alchemy which we examined; we could find no clue to the curious porcupine-like animal. The symbol c is clearly of astronomical origin, and is not often met with in later works. The MSS. are for the most part devoid of figures, and not so full of symbols as later alchemical treatises.

We have endeavored to prove (a) that no reliable date can be assigned to existing Greek MSS. on alchemy, and (b) that the accepted date is too early. Even if we could prove that a man named Zozimus, living in the fourth century, wrote treatises on alchemy, we could not use the existing MSS. for any exact purpose connected with the history of science with safety; (c) since we have no such MS. earlier than the tenth or eleventh centuries, it would be quite impossible to determine whether additions had been made during transcription.

The facts are simply these: There exist in various parts of the world Greek MSS. on alchemy, none of which are older than the tenth century. Many of these bear the names of mythical personages of Egyptian mythology, some of ancient Greek philosophers, some of people who are supposed to have lived in the fourth or fifth century, A.D. When we remember that no ancient writer makes mention of alchemy or chemistry, that the word χημεία is first used in the eleventh century, and when we further bear in mind the condition of the intellectual world in the fourth and fifth centuries, we think we may well admit that further evidence is necessary before we can assert that alchemy arose in the fourth century. Indeed we are of opinion that, in spite of all that has been written on the subject, there is no good evidence to prove that alchemy and chemistry did not originate in Arabia not long prior to the eighth century, A.D,

Chapter VII.

Latin and English MSS. on Alchemy - Sources from which the earlier Alchemists acquired knowledge - Arabic learning during the Middle Ages - Geber.

In the last chapter we discussed the Greek MSS. on alchemy, and endeavored to show, that, owing to the uncertainty of their age and the obscurity of their authorship, they are less important components of the early history of chemistry than some writers have labored to prove them.

There exist also many MSS. in Arabic and Persian on alchemy, but in all probability few of them are earlier than the 8th century. The Library of El Escorial is undoubtedly more rich in such MSS. than any existing library; but from the imperfect manner in which its treasures are cataloged, we are unable even to give a list of the more important of these treatises. The British Museum contains several Arabic MSS. on alchemy, written about the 12th century. Such of these as we have seen are devoid of drawings, and apparently also of symbols.

Early MSS. on alchemy in Latin exist in all large libraries. They contain various recipes for making the philosopher's stone, "secrets of art," copies of the inscription of the Smaragdine table, with the interpretation thereof, and an infinite amount of unintelligible nonsense. They differ in no respect from the later printed treatises on alchemy, which we shall presently discuss in detail. The matter of most of the MSS. is to be found in printed works compiled by alchemists of the fifteenth and sixteenth centuries.

One of the oldest alchemical MSS. in the British Museum is a transcript of the Speculum Secretorum of Roger Bacon, who died in 1284. It is in the Sloane Collection, and was written towards the end of the 13th century, say between 1290 and 1300.

There is no autograph MS. of Roger Bacon either in the British Museum or in the Record Office; the MS. in question was copied by an unknown man.

Alchemical MS. XIII century

Fig. 9- Alchemical MS. of the thirteenth century-British Museum.

The woodcut (Fig. 9) represents a few lines of the commencement of the MS., which reads as follows: "Incipit speculum secretorum alkimiæ. In nomine Domini Nostri Jesu Christi ad instructionem multorum circa hanc artem studere volehtium, quibus deficit copia librorum, hie libellus edatur, speculumque secretorum indicatur, idcirco quia in illo, quasi in speculo, totum secretum philosophorum et operatio eorum in hac arte, nee non et ordo operis, sensibiliter inspiciatur. Et habeant amici nostri posteri ex ejus inspectu sine tedio delec-tationem, sine obscuritate viam hoc opus aggrediendi, sine difficultate artem operandi." The translation is as follows: "In the name of our Lord Jesus Christ, for the information of the many who wish to devote themselves to the study of this art, and who lack a supply of books, this small manual is published, and is entitled the "Mirror of Secrets" seeing that in it, as in a mirror, the whole secret of philosophers and their working in this art—nay more, the process of their work—may be visibly discerned. And may our friendly descendants obtain from the perusal of it unwearied delight, a clear path for taking his work in hand, and a mode of operation unhampered by obstacles."

Among the earlier English MSS. on alchemy in the British Museum is one which, the Preface informs us, was done "at the instance and prayer of a pour creature, and to the helping of man, I, Malmedis, being at greete uneased in prisone, have thees forseide bokes hidre to itake a hand, and so I shal fynnysshe hit, to God be the laude and preisyng."

The following woodcut (Fig 10) represents a portion of this MS. relating to mercury

Fig. 10.-English MS. on Alchemy. Fifteenth century.

We must express our great indebtedness to Mr. Maunde Thompson, of the British Museum, for allowing us ready access lo the MSS.

It will be noted that mercury, together with sulphur, and the "rede stoone," is designated the producer of all metals; we also observe an allusion to the Aristotelian theory of the elements (of which an account has been given in the second chapter) in the assertion that mercury is "hotter and moyste." This MS. is in the Sloane collection, and is well preserved, and written on vellum.

Let us now turn our attention to the dogmas of the alchemists and early chemists, as set forth in the numberless printed books on the subject.

We must bear in mind at the outset that chemistry and alchemy understanding by the former, legitimate inquiry into the nature of different kinds of matter, and by the latter, the futile attempts to make gold existed side by side in the same age, often in the same person. We cannot agree with M. Hoefer when he says "La chimie du moyen âge, c'est l'alchimie," because some of the early chemists were not alchemists, and the crude processes of the one often led to the exact processes of the other. Lord Bacon, in the De Avginentis Scientiarum, has some very pertinent remarks regarding alchemy: "Credulity in arts and opinions," he remarks, "is likewise of two kinds, viz., when men give too much belief to arts themselves, or to certain authors in any art. The sciences that sway the imagination more than the reason are principally three, viz., Astrology, Natural Magic, and Alchemy. . . . Alchemy may be compared to the man who told his sons that he had left them gold, buried somewhere in his vineyard; where they by digging found no gold, but by turning up the mold about the roots of the vines, procured a plentiful vintage. So the search and endeavors to make gold have brought many useful inventions and instructive experiments to light."

The heritage which the alchemists and early chemists received from the ancients was by no means insignificant; for they possessed all the experience accumulated by the ancients in the various arts and processes which we have before described; and of theoretical matter they possessed, adopted, and prized, the theory of the transmutation of the elements proposed by Aristotle. Of works bearing upon the history of matter, they had the writings of Aristotle, Dioscorides, Lucretius, Archimedes, Hero, Vitruvius, and Pliny. Few books are quoted more often in alchemical treatises than the "Natural History" of Pliny; and we sometimes find an almost verbatim transcript of certain portions of this work. The alchemists can therefore scarcely be said to have created a science, for the science of their day is linked with that of the ancients.

When ancient learning had almost died out, and Europe was, intellectually, in a state of complete darkness, the Arabians were the most cultivated people in the world. It is to Arabia that we must look for the origin of several sciences which we are wont to attribute to other nations. The Arabians instituted universities, observatories, public libraries, and museums; they collected together all the remains of ancient learning, and through their medium the greater number of Greek and Latin authors which were read during the Middle Ages were known to Europe.

In the eighth century the Arabs had full possession of Spain, and at a somewhat later date this country-possessed the most famous universities in Europe. The Arabs, in propagating their new religion, propagated also the remains of ancient culture, which had already been introduced into Persia and Syria by the Nestorians, who had founded a school of great reputation at Odessa. Again, when Justinian closed the schools of Athens and Alexandria, many of the professors fled to Persia and Arabia, and formed new centers of learning. The works of many authors, including Aristotle, Dioscorides, and Pliny, were soon translated into Arabic and Persian, and became widely diffused.

"Ce fut," remarks M. Figuier, "ainsi que de Flnde jusqu'i l'Espagne, des rivages du Tigre jusqu'ct ceux du Guadalquivir, les livres de science se propagerent parmi des peuples qui avait deja une litterature, un philosophic religieuse, et qui n'éaient point depourvus d'imagination."

In the eighth century the University of Bagdad was founded by the Caliph Al-Mansor, and in the following century it attained a preeminent position. A large medical school was connected with it, also hospitals and laboratories. The Caliph Al-Mamoum erected an observatory in Bagdad, and an attempt was made to measure an arc of the meridian. It is said that at one time the University of Bagdad possessed more than six thousand students. In it the sciences found a home, and every scrap of ancient learning was eagerly collected and often extended. When the Arabic empire was broken up by internal dissensions into a number of small states, the University of Bagdad, losing the powerful patronage of the Caliphs, fell into decay, and soon ceased to be known. A somewhat celebrated school arose in Cairo in the tenth century, but we possess but few particulars concerning it.

We soon hear of Spain as a focus of learning. In the tenth century this was the most flourishing country in Europe, both intellectually and otherwise. The University of Cordova possessed great celebrity, and students flocked to it from all parts of the world. It contained a library of between 200,000 and 300,000 volumes, an unusually large collection of books prior to the invention of printing. The Arabians were great mathematicians and astronomers. Lalande places Mohammed-ben-Giaber (better known as Alba-tegnius) among the twenty greatest astronomers who have ever lived. Again, Alhazen wrote a treatise on optics in the eleventh century, and there were many treatises on botany and medicine. The Arabs made but little advance in anatomy however, because they were forbidden by the Koran to mutilate the human body.

After the above remarks it is almost needless to say that we must look to Arabia for the earliest treatises on alchemy and chemistry. Indeed the Arabians cultivated the latter science with success, and the first work on the subject with which we are acquainted was written by Yeber-Abou-Moussah-Djafer al-Sofi, whom we call Geber, an Arab of the eighth century. There had, no doubt, been writers on chemistry before his time, but probably not long before. We have endeavored to prove in the preceding chapter that the Greek MSS. on the "sacred art" are not trustworthy evidences of the early origin of the science; and we cannot tell from what source Geber acquired any of his knowledge. He alludes to no one by name, but we know that the Arabians collected knowledge from every source—Egyptian, Indian, Persian, Greek, and Roman. It is thought by some that Geber acquired some of his notions of chemistry from Egypt.

Several MSS. purporting to contain the writings of Geber exist in various libraries in Europe; these were translated into Latin as early as the year 1529, and into English in 1678. We have reason to believe that the Latin translation was faithfully done, if the Arabic text be not corrupt. The work consists of four treatises:—(a) Of the search for Perfection, (b) Of the Sum of Perfection, (c) Of the Invention of Verity, and (d) Of Furnaces.

Geber was acquainted with the seven metals known to the ancients, and he regarded gold, silver, copper, iron, tin, and lead, as compounds of mercury with sulphur in different proportions. Gold and silver are the most perfect metals, and are composed of the purest mercury and sulphur; the other metals consist of less pure mercury and sulphur, but may be converted into gold and silver by purifying their constituents, and uniting them in different proportions.. He describes various chemical substances, among others the following. The carbonates of potash and soda were known to Geber, and were procured from the ashes of plants. Caustic soda was procured from the carbonate by heating its solution with quicklime, as in the present day. Common salt was purified by ignition, solution, and filtration, and the solution was afterwards evaporated, and the salt crystallized out Nitrate of potash, or saltpeter, and chloride of ammonia, or sal ammoniac, were apparently common in Geber's time; as also were alum, borax, and green

Distillation

Fig 11. Distillation apparatus, from Geber's works

coppers, or protosulphate of iron. Geber procured nitric acid by distilling coppers, saltpeter, and alum, and, he used the acid for dissolving silver, and when mixed with sal ammoniac, for dissolving gold. He obtained nitrate of silver in the form of crystals, and noticed their fusibility. Various compounds of mercury are described, among others corrosive sublimate or chloride of mercury, cinnabar or sulfide of mercury, and the red oxide of mercury, in which, nearly ten centuries later, oxygen gas was discovered by Dr. Priestley. Geber also obtained sulfuric acid by distilling alum. He appears to have been acquainted with the various processes of distillation, sublimation, calcination, filtration, and many others; indeed, with almost all the processes practiced by his successors during the succeeding eight or nine centuries.

It is probable that some of the processes described by Geber were worked out in the medical schools of Arabia, and were known shortly before his time ; yet he was himself a patient worker, and often intersperses his descriptions of substances and processes with remarks on the method of experimenting, and the mode of thought most suitable for the studies which he describes. He has often been called the "Founder of Chemistry"; at least his works are the earliest with which we are acquainted, and he was venerated as Master alike by the alchemists and chemists of the Middle Ages.

Geber appears to have been acquainted with many chemical appliances. In the earliest translations of his works we find figures of various furnaces and forms of distillation apparatus; one of them, not unlike a still now in use, is represented above (Fig 11). The greater number of vessels described and figured by Baptista Porta in his treatise "De Distillationibus," published in 1609, are to be found in the first Latin translations of the works of Geber.

Chapter VIII

Avicenna - Albertus Magnus - S. Thomas Aquinas - Roger Bacon - Raymond Lulli - Arnoldus de Villâ Novâ - George Ripley - Basil Valentine.

The Schools and Colleges of Arabia soon gave evidence of their value by the development of several considerable geniuses, whose works formed the textbooks of Europe during a portion of the Middle Ages. Prominent among these learned Arabians was Ali-ben-Sina, or Avicenna, who was born in 980, in the neighborhood of Shiraz. His abilities were considerable, and no pains were spared in his education; as a boy he read the Almagestum of Ptolemy, the Geometry of Euclid, and the Philosophy of Aristotle, and later in life he studied medicine with great success. We are told indeed that at the age of sixteen he was an eminent physician, and that at eighteen he cured a caliph of some grave disorder, and was hence promoted to great honor.

Avicenna is best known by his celebrated "Canons", which were translated at an early date into Latin, and often printed under the title of "Canones Medicinae". This work has been translated into the languages of all civilized countries, and for no less than six centuries was the standard medical treatise of the world.

Avicenna also wrote on alchemy and on chemistry. If the works attributed to him are genuine, he appears to have adopted the Aristotelian theory of the four mutually convertible elements. He speaks of air as the aliment of fire, and of the metals as compounds of a humid substance and an earthy substance. This last idea evidently arose from the observation of the calcination of metals. It was well known that if certain metals, such as lead and tin, are heated for a length of time in the air, they are converted into a powdery substance or calx, and it was long before it was proved that this calx is not the metal from which one of its constituents has been expelled by fire; but, on the other hand, the metal combined with another substance. Avicenna divides all minerals into four classes: viz. (1) Infusible minerals; (2) Minerals which are fusible and malleable, that is, metals; (3) Sulfurous minerals; and (4) Salts. He noticed that mercury can, by heat, be caused to unite with sulphur and produce a solid body, having different properties from its constituents.

Avicenna was largely indebted for his knowledge to Alfarabi and to Rhazes. The latter wrote on medicine, and was one of the first to introduce substances formed artificially by chemical means into medicine.

Turning now our attention to European alchemists, we meet at the outset with the name of Albertus Magnus (b. 1193, d. 1282), who became Bishop of Ratisbon in 1259. Various works on alchemy are attributed to him; he wrote on the Philosopher's Stone, on the origin of metals, and on minerals; and he has described at some length various chemical operations, such as sublimation and distillation, and various forms of apparatus, such as aludels, alembics, and water-baths. He followed Geber in the belief that metals are composed of sulphur and mercury, and that different metals are produced by different combinations, and to some extent by the variations in the purity, of these substances.

Albertus Magnus employs the term affinity (affinity) to designate the cause of the combination of sulphur with silver and other metals; in this precise sense, applied to all cases of chemical combination, the term is used in the present day. He also speaks of sulfate of iron as vitriol, a name which it long retained. He describes the preparation of nitric acid, its principal effects upon certain metals, and its utility for separating silver from gold, inasmuch as it will dissolve the former and not the latter. Cinnabar, or sulfide of mercury, had long been known and used as a source of mercury; Albertus proved that it consists of sulphur and mercury by preparing it artificially, by subliming sulphur with mercury.

Albertus was not alone learned in alchemy; he was a profound theologian, a scholar, an astronomer, a physician, and some said an adept in magic and necromancy. He embodied his wisdom in twenty-one folios, which were published in a collected form in 1651. M. Lenglet Dufresnoy, in his "Histoire de la Philosophic Hermétique," has mentioned several magical operations gravely attributed to Albertus Magnus by various writers. The most noticeable piece of magic was the sudden transformation of a winter's day into glowing summer: "Horridam hyemem," says Trithemius", in florigeram fructiferamque vertit. It is said that once during a very severe winter, he invited Count William of Holland, when he was passing through Cologne, to a feast. The Count, on his arrival with a considerable retinue, was surprised to find the feast spread in the garden, in which there was a depth of several feet of snow; and this treatment so angered him that he remounted his horse and prepared at once to leave his inhospitable host.

"Then the monk falling on his knees besought The Count to sit one moment at the board. He having done so, a most wondrous change Passed on the instant over all around. The dark clouds floated off and left a sky Intensely blue, an air exceeding clear; The sun shone brightly, and the warm south wind Laved their pale cheeks and warmed them into life. They sit on greenest grass, the snow is gone, Sweet flowers bloom beneath their very feet, Ripe peaches blush upon the garden wall, And orange blossoms scent the humid air. A swarm of insect life on droning wing Is floating up above them in the breeze. The voice of birds is heard; the cooing dove Speaks softly to her mate; the nightingale Trills a sweet lay, half hidden in the leaves. All nature is most joyous in her garb Of brightest summer day, and all things seem To glory in the flood of warmth and light."

Upon this, the Count expressed considerable astonishment, as, although he had heard a good deal of the magical powers of his host, he was quite unprepared to find him capable of changing the seasons. As soon as the feast was ended, Albertus repeated a magical formula—

"Now snow obscures the air, the flowers fade, The trees are torn by pitiless strong winds, And weep their shriveled fruit upon the earth: All sound of life is gone, a roar of elements Succeeds the plaintive quavering of the leaves. The birds fall dead to earth, and the dark air Betokens fearful tempests yet to come."

So the Count and his retinue rush off into the house to warm themselves, and thus ends the feast of Albertus Magnus. Some will have it that the story-alludes to a winter garden, which had been devised by Albertus for the preservation of rare plants, and which was unknown at that time. Middle Age books on science abound with such stories, and the belief in them was almost universal, as it well might be in an age in which the power of witches and wizards was acknowledged, and the raising of the dead was an admitted possibility. Brücker (Institutiones Historice Philosophicce) says

:— a "Quae enim de ejus convivio magico narrantur, merito inter inficeti seculi fabulas referuntur, quae ex ignorantia rerum naturalium eo tempore crassissima et Alberti mirabili rerum physicarum cognitione prodierunt."

In the church of S. Andreas in Cologne they show to this day the shrine and relics of Albertus - the accomplished churchman, scholar, magician and alchemist, of whom Trithemius says, "Magnus in Magia Naturali, major in Philosophia, maximus in Theologia."

Albertus had for his pupil the "angelic doctor," S. Thomas Aquinas (b. 1225, d. 1274), who was a great alchemist, and who wrote a treatise called "The most secret Treasure of Alchemy," together with some other works on the subject, which are equally obscure and unintelligible. He wrote also on the artificial preparation of gems, by fusing glass with certain substances, like oxide of copper, to communicate different colors; he mentions that if copper be heated with white arsenic, the former becomes white, something like silver. According to some, S. Thomas Aquinas was the first to employ the term amalgam, to designate a compound of any metal with mercury. S. Thomas Aquinas was, like his master, a magician. We are told that between them they constructed- a brazen statue, which Albertus animated with his elixir vitæ. It was useful as a domestic servant, but very talkative and noisy; nor could they cure it of this propensity. It happened one day that S. Thomas, who was a mathematician, was deeply engaged in a problem, but was continually interrupted by the talking statue; at length in a rage he seized a hammer and smashed it to atoms, to the great regret of his master.

Our great countryman Roger Bacon (b. 1214) also suffered from a charge of magic, and during his residence in Oxford was severely persecuted in consequence. He replied to the charges made against him by the admirable treatise "De nullitate Magiæ," and in it clearly showed that what his contemporaries mistook for the work of spirits, was in good sooth due to the ordinary operations of Nature. In this work he speaks of gunpowder, although somewhat obscurely. "Mix," says he, "together saltpeter luru vopo vir con utriet, and you can make thunder and lightning, if you know the method of mixing them." Elsewhere he says,"a small quantity of matter properly manufactured, and not larger than one's thumb, may be made to produce a horrible noise and sudden flash of light." The third constituent of gunpowder is designated under the anagram luru vopo vir con utriet, for it was dangerous in those days to speak too plainly; indeed Bacon tells us that he adopted an obscure style both on account of the example of other writers, and of propriety, and also on account of the dangers of plain speaking. According to some writers, the following passage is to be found in Bacon's writings: "Sed tamen salis petrae, luru mone cap ubre, et sulphuris, et sic facies tonitrum si scias artificium." Thus the saltpeter and the sulphur are directly designated, while the anagram luru mone cap ubre is convertible into carbonum pulvere, the remaining constituent powdered charcoal. It is improbable that Roger Bacon invented gunpowder, although he was the first to know of its properties in England; he probably procured the knowledge from an Arabic source. Gunpowder was first used by the English at the battle of Crecy in 1346, sixty-one years after the death of Bacon; at this time it was apparently unknown to other European nations.

Roger Bacon is believed to have been far in advance of his times in all matters of science. To him has been attributed the invention of the telescope and Camera obscura, and several discoveries of a later date. The evidence is less conclusive than one could wish, but enough remains in his writings to prove that he was a very learned man and profound thinker. His "Opus Majus" clearly proves that he fully recognized the value of the experimental method, and of the inductive philosophy afterwards so ably advocated by his namesake Francis Bacon. Roger Bacon wrote largely on alchemy. Many of the alchemical MSS. in the British Museum are transcripts of portions of his works, among the more celebrated of which we may mention the " Medulla Alchymiæ", "Secretum Secretorum", and "Speculum Secretorum." He collected together the principal alchemical facts of his predecessors, and appears in many matters to have closely followed Geber. Bacon describes the distillation of organic substances, and alludes to the inflammability of the evolved gases. He proved that air is the food of fire by burning a lamp in a closed vessel.

Raymond Lulli (b. 1235) is by some asserted to have been a pupil of Roger Bacon. He was a voluminous writer on alchemy, his most celebrated treatise being his "Ultimum Testamentum." He also wrote on transmutation, on the Philosopher's Stone, and on magic. Lulli does not appear to have added to the chemical knowledge of his predecessors; he followed Geber closely, and was well acquainted with the processes and compounds which he describes. He describes alcohol under the names of aqua vitas ardens, and argentum vivunt vegetabile and was in the habit of rendering it anhydrous by allowing it to stand in contact with dry carbonate of potassium. He was also acquainted with ammonia.

Whatever Lulli's knowledge may have been, he obtained great reputation as a successful alchemist. He asserts in his "Ultimum Testamentum" that he converted fifty thousand pounds weight of base metals into gold. He is said to have been employed by one of the Edwards to make gold, and to have furnished His Majesty with six millions of money. Dickenson tells us that Lulli had a laboratory in Westminster Abbey, in which, after his departure, a quantity of gold dust was found.

Of the general tone and character of alchemical writings we shall speak more fully in the next chapter. Of the professors of the art little more need be said; a long list of names might be given, but it would be found that they did little to develop what afterwards became the science of chemistry. Let us glance at the work of a few of the remaining alchemists. Arnoldus de Villa Nova (b. 1240) was a great alchemist and physician, and the author of many works on the subject. His "Rosarius Philosophorum" purported to contain a key to all alchemical operations. He followed Geber closely. He considered a solution of gold the most perfect medicine, and we usually find that such solution was recommended by alchemists as a necessary constituent of the elixir vitæ, and essential for the work of transmutation. In Fig. 12 the solution of gold in the flask is represented by the sun emitting rays. The simple disc of the sun is the more common symbol for gold.

Arnoldus also distilled various oils and essences. He contended that sulphur, arsenic, mercury, and sal ammoniac—all volatile bodies be it noted—are the souls of metals, and are given off during calcination. He also affirmed that silver is intermediate between mercury and other metals, just as the soul is intermediate between the spirit and the body. Arnoldus is said to have had for his pupil Pope John XXII., an accomplished alchemist, who left at his death eighteen millions of florins, which the alchemists fondly cite as a proof of the possibility of transmutation.

Fig. 12.—An alchemist hermetically sealing a flask containing a solution of gold.

Hermetically Sealing

Fig.12-An Alchemist hermetically Sealing a Flask containing a solution of gold.

Our countryman, George Ripley, Canon of Bridlington in Yorkshire (b. about 1460), wrote a poem on alchemy, and passed fur a successful disciple of the art, but we cannot point to a new fact which he elucidated. He divided all chemical operations into twelve processes—calcination, dissolution, separation, conjunction, putrefaction, congelation, cibation, sublimation, fermentation, exaltation, multiplication, and projection. Several MS. copies of his poem exist in the British Museum, bound up with copies of the works of Roger Bacon and earlier writers. Here is a specimen of his rugged rhymes:

Fig. 13.—Alchemical representation of processes.

Processes

Fig.13-Alchemical Representation of Processes

The first chapter shall be of natural Calcination;

The second of Dystolution, secret and phylosophycall;

The third of out elementall Separation;

The fourth of Conjunction matrimoniall;

The fyfth of Putrefaction then followe shall:

Of Congelation Alhyficative shall be the sixt.

Then of Cybation, the seventh shall follow next.

One of the most celebrated of the alchemists was Basil Valentine, who was born at Erfurt in 1394.

According to Olaus Borrichious, his works were accidentally discovered in the wall of a church at Erfurt many years after his death. A thunderbolt struck the church and exposed to view the long-lost alchemistical treasures. Basil Valentine was the author of many treatises, the most important being his "Currus Triumphalis Antimonii," in which he discusses the properties of antimony and of some of its compounds. He regarded the metals as compounds of salt, sulphur, and mercury; and he was acquainted with many metallic compounds, among others nitrate of mercury, sulfide of arsenic, red oxide of mercury, chloride of iron, sulfate of iron, fulminating gold, carbonate of lead, acetate of lead, and the oxides of lead. He was aware that iron precipitates copper from solution, and that solution of potash precipitates iron from solution. He was well acquainted with the preparation of nitric and sulfuric acids, and used them for various purposes of dissolution. In order to obtain nitric acid he distilled powdered earthenware with nitre, or equal weights of nitre and green vitriol, or nitre with finely powdered flints. He obtained fuming sulfuric acid by distilling green vitriol, after the manner still practiced at Nordhausen and elsewhere. Basil Valentine wrote very obscurely and was fond of symbolical designs. Figures 13 and 14 are taken from his works, and represent various processes imperfectly described. Thus the lion in Fig. 13 would represent a solution of a metal, the serpent another solution, or perhaps the serpent a metal, and the lion devouring it a solvent; the sun and moon are watching the operation, and the symbol of mercury appears between two roses.

FIG. 14.—Alchemical representation of processes.

Operations

Fig. 14 represents some operation which is thus described by the principal figure; "I am an old, infirm, debilitated man, my soul and spirit (represented by the two boy-headed birds above his head) leave me, and I assimilate the black crow. In my body are found salt, sulphur, and mercury." This may possibly refer to the solution of gold in aqua regia; it loses its metallic nature, its solidity and luster, and assimilates the acid; but one may conjecture in vain concerning the enigmatically devices in which some of the alchemists took so much delight, and which they often employed, like Roger Bacon's anagram, to conceal the full significance of their operations or discoveries.The following extract, which treats of the generation of metals, will show the style of Basil Valentine's writing: "Therefore think most diligently about this; often bear in mind, observe, and comprehend that all minerals and metals together in the same time, and after the same fashion, and of one and the same principal matter are produced and generated. That matter is no other than a mere vapor, which is extracted from the elementary earth by the superior stars or by a sidereal hot infusion, with an airy sulfurous property, descending upon inferiors, so acts and operates as in those metals and minerals is implanted spiritually and invisibly a certain power and virtue, which fume afterwards resolves in the earth into a certain water from which mineral water all metals are thenceforth generated and ripened to their perfection, and thence proceeds this or that metal or mineral according as one of the three principles acquires dominion, and they have much or little of sulphur and salt, or an unequal mixture of them; whence some metals are fixed, that is constant or stable; some volatile and easily mutable, as is seen in gold, silver, copper, iron, lead, and tin."

Now this is by no means the most obscure piece of alchemical writing with which we shall come in contact.

Chapter IX.

General Character of Alchemy and the Alchemists - The "Pretiosa Margarita Novella" - An Alchemistical Allegory - Alchemical Symbols - Paracelsus - Libaviuss.

What manner of men were the alchemists? How did they preserve, cultivate, and transmit the wonderful delusions of their creed? We have endeavored in a former chapter to show that the idea of transmutation arose from the old Greek idea of the conversion of one element into another; and the belief in the possibility of transmutation once admitted, the pursuit of the alchemist would naturally follow in a mystical and credulous age. As to the men themselves, their character was twofold; for there was your alchemist proper, your true enthusiast, your ardent, persevering worker, who believed heart and soul that gold could be made, and that by long search or close study of the works of his predecessors, he could find the Philosopher's Stone; and there was your knavish alchemist, a man who had wits enough to perceive that the search was futile, and impudence enough to dupe more credulous people than himself and wheedle their fortunes out of them on pretense of returning it tenfold in the shape of a recipe for converting lead into gold. These last we may dismiss at once. They abounded during the Middle Ages, and found easy dupes, whom they deceived by the most shallow tricks, as by placing a piece of gold in the crucible of transmutation together with volatile substances, and after many processes and much heating, they would show the little button of metal which had all along been present.

Of the true alchemist we have many pictures. The alchemist, the astrologer, the mystic, the wizard, were men of the same stamp. They often practiced the same arts side by side. The same habit and attitude of thought belonged to one and to all, and became all equally well. Take the dreamy, maudlin, semi-maniacal Althotas, who has been described so well by Dumas: "An old man, with gray eyes, a hooked nose, and trembling but busy hands. He was half-buried in a great chair, and turned with his right hand the leaves of a parchment manuscript." Note also his intense abstraction, his forgetfulness of the hour, the day, the year, the age, the country; his absolute and intense selfishness and absorption, the concentration of the whole powers of his soul upon his one object. Or let us look at Victor Hugo's Archidiacre de St. Josas, in his search for the unseen, the unknown, and the altogether uncanny; the bitterness of his soul, his passionate musings, his conjurations and invocations in an unknown tongue; his own self, that wonderful mixture of theologian, scholar mystic, perhaps not much unlike the divine S. Thomas Aquinas himself.

Listen to his musings: "Yes, so Manon said, and Zoroaster taught: the sun is born of fire, the moon of the sun; fire is the soul of the universe; its elementary particles are diffused and in constant flow throughout the world, by an infinite number of channels. At the points where these currents cross each other in the heavens they produce light, at their points of intersection they produce gold. Light! gold! the same thing; fire in its concrete state.... What! this light that bathes my hand is gold? The first the particles dilated according to a certain law, the second the same particles condensed according to another law! . . . For some time, said he, with a bitter smile, I have failed in all my experiments; one idea possesses me, and scorches my brain like a seal of fire. I have not so much as been able to discover the secret of Cassiodorus, whose lamp burned without wick or oil a thing simple enough in itself." If we peep into Dom Claude's cell, we are introduced to a typical alchemists laboratory—a gloomy, dimly-lighted place, full of strange vessels, and furnaces, and melting-pots, spheres, and portions of skeletons hanging from the ceiling; the floor littered with stone bottles, pans, charcoal, aludels, and alembics, great parchment books covered with hieroglyphics; the bellows with its motto Spit a, Spera; the hour-glass, the astrolabe, and over all cobwebs, and dust, and ashes. The walls covered with various aphorisms of the brotherhood; legends and memorials in many tongues; passages from the Smaragdine Table of Hermes Trismegistus; and looming out from all in great capitals, ANAρKH. Yet once again, look at Faust, as depicted by Rembrandt; or Terriers unknown alchemist, if you wish for an alchemical interior.

But the hard-working and enthusiastic alchemist did not always follow the ideal of the novelist and artist; he often degenerated into a "dirty soaking fellow," who lost what little learning he ever had by concentrating his mind on the one dominant topic, until it excluded every other idea and aspiration; then the pursuit became all-absorbing, and the disciple of the art a mere driveling monomaniac

We will now look at one of the books which were cherished by the alchemists. Here is a little vellum-covered Aldus: date 1546. Paracelsus had been dead five years, and Cornelius Agrippa twelve years; Dr. Dee and Oswald Crollius were flourishing; Van Helmont and a host of known alchemists were unborn. Our little volume, full of quaint musings of a bygone age, has outlived them all, and yet it never drank of the elixir vitæ although it pretended to teach others how to make it, and the Philosopher's Stone into the bargain. "Pretiosa Margarita Novella de Thesauro, ac pretiosissima Philosophorum Lapide " is the title; published with the sanction of Paul III., Pontifex Maximus, whose successor, be it remembered, established the "Index Expurgatorius," and might possibly have prohibited this Precious Pearl of alchemy. The title-page tells us that it contains the methods of the "divine art," as given by Arnoldus de Villa Nova, Raymond Lulli, Albertus Magnus, Michael Scotus, and others, now first collected together by Janus Lacinius. The vellum cover is well thumbed, and in one place worn through, perhaps by contact with a hot iron on an alchemist's furnace-table, or by much use. There are no MS. notes, but on the title-page is the autograph of Sir E. Koby, or Hoby, and a favorite maxim, the first word of which—Fato—is alone legible. The date of the writing is perhaps 1580-90. Some initial letters of the text have been plainly illuminated in red, by a loving hand; they were copied from a Bible transcribed at Lyons in 1326.

FIG. 15.—Allegorical representation of transmutation.

Transmutation

Fig.15-Allegorical Representation of Transmutation

As to the contents, we have firstly an opening address by Janus Lacinius ; then certain definitions of form, matter, element, color, &c; next, symbolic representations of the generation of the metals, and after this a woodcut representing the transmutation of the elements according to the dogmas of Aristotle. (See Chapter I Fig 1.) After this we find the whole course of transmutation set forth pictorially and allegorically, as under. A king (see Fig. 15), crowned with a diadem, sits on high, holding a scepter in his hand. His son, together with his five servants, beseech him on bended knees to divide his kingdom between them. To this the king answers nothing. Whereupon the son, at the instigation of the servants, kills the king and collects his blood. He then digs a pit, into which he places the dead body, but at the same time falls in himself, and is prevented from getting out by some external agency. Then the bodies of both father and son putrefy in the pit Afterwards their bones are removed, and divided into nine parts, and an angel is sent to collect them. The servants now pray that the king may be restored to them, and an angel vivifies the bones. Then the king rises from his tomb, having become all spirit, altogether heavenly and powerful to make his servants kings. Finally he gives them each a golden crown, and makes them kings (Fig. 16).

FIG. 16.—Allegorical representation of transmutation.

Transmutation done

Fig.16-Allegorical Representation of Transmutation

It is difficult to follow this from beginning to end, but there can be no doubt that the king signifies gold, his son, mercury, and his five servants the five remaining metals then known, viz. iron, copper, lead, tin, and silver. They pray to have the kingdom divided among them, that is to be converted into gold; the son kills the father, viz. the mercury forms an amalgam with gold. The other operations allude to various solutions, ignitions, and other chemical processes. The pit is a furnace; putrefaction means reaction or mutual alteration of parts. At last the Philosopher's Stone is found; the gold, after these varied changes becomes able to transmute the other metals into its own substance. At the end some rugged hexameters and pentameters warn the fraudulent, the avaricious, and the sacrilegious man that he is not to put his hands to the work, but to leave it for the wise and the righteous, and the man who is able rightly to know the causes of things.

After this allegory we have some remarks concerning the Treasure, and the Philosopher's Stone, and the Secret of all Secrets, and the Gift of God. This is followed by a number of arguments against alchemy, and of course overwhelming arguments in favor of it. Among those who are quoted as alchemists are Plato, Pythagoras, Anaxagoras, Democritus, Aristotle, Morienus, Empedocles, and then, with a delightful disregard of age or country, we read, "Abohaby, Abinceni, Homerus, Ptolemaeus, Virgilius, Ovidius." Then digressions on the difficulties of the art, the unity of the art, the art natural and divine; a slight history of the art, in which it is traced back to Adam, although Enoch and Hermes Trismegistus are mentioned as possible founders. A treatise to prove that this art is more certain than other sciences; on the errors of operation; on the principles of the metals; on sulphur; on the nature of gold and silver; and many general remarks on all alchemical subjects. These are the teachings which the "Pretiosa Margarita Novella " pours at the feet of the wise among mankind, by the aid of Paulus Manutius, bearing his father's name of Aldus, and by the grace of the Venetian Senate.

Many attempts were made by the alchemists to explain the origin of the metals. Some regarded them as natural compounds of sulphur and mercury; others affirmed that the power of the sun acting upon and within the earth produced them, and that gold was in truth condensed sunbeams. Many believed that metals grew like vegetables; indeed it was customary to close mines from time to time to allow them to grow again. Basil Valentine, as we have seen, regarded them as condensations of a " mere vapor into a certain water, "by which latter we suspect he meant mercury. Perhaps the most absurd account of the origin of certain things is given by Paracelsus in his treatise, "De Natura Rerum," in the following words, which will show also how utterly nonsensical and unintelligible alchemical language could be, and for that matter very generally was. "The life of metals," he writes, "is a secret fatness; ... of salts, the spirit of aquafortis; . . . of pearls, their splendour; . . . of marcasites and antimony, a tinging metalline spirit; . . . of arsenics, a mineral and coagulated poison . . . The life of all men is nothing else but an astral balsam, a balsamic impression, and a celestial invisible fire, an included air and a tinging spirit of salt. I cannot name it more plainly, although it is set out by many names."

The peculiarly secret and mystical language which the alchemists adopted was intended to prevent the vulgar from acquiring the results of their long-continued labors. Their language purported to be intelligible to the true adept; but as a rule the alchemists of one age gave various interpretations to one and the same secret communicated by their predecessors. Long recipes for the preparation of the Philosopher's Stone exist, which the authors have generously (as they tell us) given to the world, after much labor, for the benefit of their fellow-men. The obscurity of the science was increased by the multiplication of symbols; the presence of which in alchemy clearly points to its connection with astrology and the sister sciences. In time alchemical symbols multiplied almost as much as astrological symbols. In an Italian MS. of the early part of the seventeenth century which we have before us, mercury is represented by 22 distinct symbols, and 33 names, many of which are of distinctly Arabic origin: such as Chaibach, Azach, Jhumech, Caiban. Lead is represented by the symbols in Fig. 17, and in addition to its ordinary alchemical names, is called Okamar, Syrades, Malochim, and others. The designation of substances as "the green lion," "the flying eagle," "the serpent," "the black crow," and so on, also led to considerable confusion. Both names and symbols were used in a somewhat arbitrary fashion.

FIG. 17.—Symbols of lead from Italian MS. of the seventeenth century.

Lead

Fig. 17.—Symbols of lead from Italian MS. of the seventeenth century.

It is somewhat strange to think that alchemy should have once received the serious attention of the legislature in this country. In 1404 the making of gold and silver was forbidden by Act of Parliament. It was imagined that an alchemist might succeed in his pursuit, and would then become too powerful for the State. Fifty years later Henry VI. granted several patents to people who thought they had discovered the Philosopher's Stone; and ultimately a commission of ten learned men was appointed by the King to determine if the transmutation of metals into gold were a possibility. We must now leave the subject of alchemy. Those who desire to study it more deeply will find a great mass of matter in the "Bibliotheca Chemica Curiosa" of Mangetus; but if they will take our advice, they will not waste much time in studying the history and progress of a futile and false art.

Fig. 18.—Designs from Mangetus (*Bibliotheca Chemica Curiosa*).

Magentus

Fig.18-Designs from Mangetus (Bibliotheca Chemika Curiosa)

With Paracelsus (b. 1493 d. 1541) a somewhat new phase of the science of chemistry appeared. By pointing out the value of chemistry as an adjunct to medicine, he caused a number of persons to turn their attention to the subject, and to endeavor to ascertain the properties of various compounds. Thus he helped to withdraw men from the pursuit of alchemy, by asserting that the knowledge of the composition of bodies, which had necessarily been forwarded by alchemy, was of importance to the human race, for the better prevention and curing of their ills. In the way of discovery or research, Paracelsus did little. He mentions zinc and bismuth, and associates them with metallic bodies, and he makes considerable use of several compounds of mercury, and of sal ammoniac. Paracelsus compares the alchemist of his day with the physician, and speaks of the former in the following terms: - For they are not given to idleness, nor go in a proud habit, or plush and velvet garments, often showing their rings upon their ringers, or wearing swords with silver hilts by their sides, or fine and gay gloves upon their hands, but diligently follow their labors, sweating whole days and nights by their furnaces. They do not spend their time abroad for recreation, but take delight in their laboratory. They wear leather garments with a pouch, and an apron herewith they wipe their hands. They put their fingers among coals, into clay, and filth, not into gold rings. They are sooty and black like smiths and colliers, and do not pride themselves upon clean and beautiful faces."

Among the Paracelsians we find Oswald Crollius, who mentions chloride of silver under the long-retained name of luna cornea, or horn-silver, from its peculiar horny appearance and texture after fusion. He was also acquainted with fulminating gold.

The name of Andrew Libavius (died 1616) deserves mention, because he sought to free chemistry from the mazes of alchemy and mysticism in which it was involved. In this he to some extent succeeded: and he appears also to have been a patient worker in the field of the science which he did so much to promote. He discovered the per chloride of tin, which is even now called fuming liquor of Libavius; he also proved that the acid (sulfuric acid) procured by distilling alum and sulfate of iron, is the same as that prepared by burning sulphur with saltpeter Libavius was great at the making of artificial gems, and was able to imitate almost any precious stone by coloring glass with various metallic oxides.

Chapter X.

Early Ideas concerning the process of Combustion - Association of Nitre with the Air, so far as the part they play in Combustion is concerned - Hookas Theory of Combustion - Mayow's Experiments - Early Pneumatic Chemistry - Proof of the Analogy existing between Combustion and Respiration.

As in the history of matter we find molecules grouping themselves around a common center or a common line, thus constituting crystalline bodies, so in the history of sciences and of nations we may often observe well-defined axes, about which the facts of particular epochs congregate. Such axes are to be found in the history of chemistry. At the particular period of which we now write, the facts of the science mainly grouped themselves around theories connected with combustion, which involved as collateral matters conceptions regarding the nature of calcination and of the air.

Combustion was, and still is, the most prominent exhibition of chemical force, with which man ordinarily comes into contact. It is a purely chemical action—the union of dissimilar bodies under the influence of chemical affinity, attended by the evolution of light and heat. Many attempts were made to explain its cause. Fire, in common with earth, air, and water, as we have before seen, was regarded as an element, till almost within our own memory. Epicurus regarded heat as a congeries of minute spherical particles possessing rapid motion, and readily insinuating themselves into the densest bodies. Fire was simply an intense form of heat. Cardanus speaks of flame as aer accensus, and of fire as heat immensely augmented.

During the Middle Ages the existence of two kinds of fire was admitted— the one pure celestial fire "subtilis ignis" "cælestis ignis" the principle or essence of fire; the other "gross earthly fire," or "mundane fire." The latter was the materia (Latin), the former the forms. Celestial fire became mundane fire when it was associated with combustible bodies, that is, in ordinary combustion. Seneca tells us that the Egyptians divided each element into an active and a passive form; fire became active flame which burns, and comparatively passive warmth and light. The elemental nature of fire was not universally admitted during the Middle Ages; thus Francis Bacon asserts, in the "Novum Organum," that fire is "merely compounded of the conjunction of light and heat in any substance," and he defines heat as a rapid motion of material particles. Athanasius Kircher, in his ponderous treatise, "Ars Magna Lucis et Umbrae," affirms that fire is air which is caused to glow by the violent collision of bodies, by which means combustible bodies become flame. At an early date it was observed that fire cannot exist without air; the experiment of burning a candle in a closed Vessel was well known. Some affirmed that "air is the food of fire," some that "air nourishes fire." The influence of a blast of air upon fire was well recognized; we have seen that bellows were known at a very early date. When nitre (another term for saltpeter) - which for many centuries was one of the most important bodies in chemistry - came to be known, it was soon noticed that it produces intense ignition; that, in fact, to direct a blast of air upon a red-hot coal, or to throw some nitre upon it, produced the same result, viz. greatly augmented combustion.

Hence arose the idea that nitre and the air are in some way connected, for "things which are equal to the same are equal to each other". This association of ideas may seem crude to us now, yet we must remember that nitre produces rapid combustion simply because it contains a great quantity of that constituent of the air, oxygen gas, which ordinarily produces combustion. Thus the old natural philosophers, wandering in the dim twilight of experimental knowledge, were not so far wrong in their supposition. The idea mentioned above was very prevalent two centuries ago: Robert Boyle speaks of the presence of a "volatile nitre" in the air; Lord Bacon says that nitre contains a "volatile, crude, and windy spirit"; Clark attributed thunder and lightning to the presence of nitre in the air; Gassendi imagined that minute particles of nitre are diffused throughout the atmosphere. When we heat lead or tin in a current of air, these metals are respectively converted into a powder, or calx and calcination was one of the most important processes in old chemistry. Calcination seemed to be due more or less directly to the air; and metals could also be calcined by heating them with nitre, or with the spirit of nitre, nitric acid; hence arose another bond of connection between nitre and the air; at least, they had something in common. Lemery in his Cour de Chimie, published in 1675, affirms that the acid of nitre contains a number of "corpuscules igntes" locked up in it, and he defines these latter as "a subtle matter, which having been thrown into a very rapid motion, still retains the power of moving with impetuosity, even when it is enclosed in grosser matter; and when it finds some bodies which by their texture or figure are apt to be put into motion, it drives them about so strongly that, their parts rubbing violently against each other, heat is thereby produced."

Thus recognizing the causes which had led to the association of the air with nitre, at least so far as they are both concerned in the production of combustion, we are prepared to examine Robert Hooke's theory of combustion. The announcement of this theory marks an important epoch in the history of chemistry; it was the first chemical theory worthy of the name, and it gave a far more just and accurate explanation of combustion than the crude and over - belauded theory of Phlogiston, of Beccher and Stahl. Hooke's theory was, moreover, founded upon experiment, and although unfortunately he does not describe the experiments, we see at a glance that it could not have been constructed without such means. "This hypothesis" he writes, "I have endeavored to raise from an infinity of observations and experiments", and all who know Hooke's writings are well aware how good an experimenter he was. The theory was published in 1665 in Hooke's Micrographia; it is there found (Observation 16) buried in a mass of irrelevant matter, and to this cause may, perhaps, to some extent be attributed the fact that it has been so little recognized and known. The theory is stated in twelve Propositions, the principal of which,are as' follows:

1. "That the air is the universal solvent of all sulfurous bodies."

    Sulfur was long regarded as the type of combustible bodies, on account of its ready inflammability ; some even derive the name from salt wvp, the salt of fire. By sulfurous bodies, Hooke simply meant combustible bodies, viz. bodies that can burn in a supporter of combustion. By air being the "universal solvent", he meant that through the agency of air combustible bodies are caused to become transformed into similarly invisible substances. For instance, we burn a pound of wood, and a few grains of ash remain, the rest has disappeared into air; as we say now, it has been converted into carbonic anhydrous gas; as Hooke said then, it has been dissolved by the air.

2. "That this action it (the air) performs not until the body be sufficiently heated."

In more modern phraseology, every combustible possesses its special igniting point—phosphorus 92 deg. F, sulphur 482 deg. F., and so on.

3. "That this action of dissolution produces or generates a very great heat, and that which we call Fire"
4. "That this action is performed with so great a violence, and does so minutely act, and rapidly agitate the smallest parts of the combustible matter, that it produces in the diaphanous medium of the air the action, or pulse, of Light!"

This would seem to indicate that Hooke considered light to be an intensified form of heat, and to be generated in the same manner, and to be a kind of very rapid motion.

5. "That the dissolution of sulfurous bodies is made by a substance inherent and mixed with the air, that is like, if not the very same with, that which is fixed in saltpeter"

Hooke had evidently traced the connection between certain actions produced by the air and by saltpeter or nitre; and he says it may be readily demonstrated that combustion is effected by that constituent of the air which is fixed in saltpeter This is a remarkable assertion, because oxygen gas was not discovered until more than a century after the proposition of Hooke's theory; and we now know that nitre contains "fixed in it the same substance — oxygen gas — which causes air to " dissolve " combustible bodies. It is probable that the connection between air and nitre may have been rendered the more probable in the minds of Hooke and his contemporaries by the knowledge that gunpowder will burn in a space devoid of air; thus, if sulphur and charcoal burn in air, and consume air in burning, and if nitre will cause them to burn out of contact with air, it would surely appear that nitre must contain air, or one of its components.

6. "That the dissolving parts of the air are but few..... whereas saltpeter is a menstruum, when melted and red hot, that abounds more with these solvent particles, and therefore as a small quantity of it will dissolve a great sulfurous body, so will the dissolution be very quick and violent."

It was well known that if a piece of rod-hot charcoal be thrown into melted nitre, it is consumed with great rapidity, while in the air it burns with far less readiness; hence Hooke infers that that particular component of air which causes it to support combustion exists in a condensed form in saltpeter He also remarks, that if air be violently forced upon a piece of ignited charcoal by bellows it may be made to burn almost as rapidly as in melted nitre.

7. "It seems reasonable to think that there is no such thing as an element of fire .... but that that shining transient body called flame is nothing else but a mixture of air and the volatile sulfurous parts of dissoluble or combustible bodies."

Hooke asserts that this theory had been worked out by him several years earlier, and had been well supported by experimental means: he says, moreover, that he has here "only time to hint an hypothesis, which, if God permit me life and opportunity, I may elsewhere prosecute, improve, and publish." This he never did; but a young Oxford physician named John Mayow (b. 1645, d. 1679) eagerly accepted the theory, and adduced many experiments in support of it.

Perhaps Mayow may have worked with Hooke during his residence in Oxford, and may have helped to adduce verification's of the then half-formed theory.

Mayow's experiments are contained in a treatise entitled Tractattis Quinque Medico-Physics quorum primus agit de Sal-nitro et Spiritu Nitro-aereo, Secun-dus de Respiratione" .... Oxonii, 1674." The book is altogether important, because the experiments which it contains form the basis of pneumatic chemistry, that is, the chemistry of gaseous bodies; it is also distinguished by accurate reasoning and well-founded generalizations Had it been better known, it can scarcely be doubted that the discovery of oxygen, and of various gases made a century ago, would have been forestalled by many years.

Fig. 19 —John Mayow.
(From his "Tractatus Quinque Medico-Physici, 1674.")

John Mayow

Fig. 19-John Mayow

Mayow calls the "dissolving parts" of the air and of nitre, which we now call oxygen gas, by the several names of nitre-air, fire-air, and nitro-ærial spirit. Air does not consist wholly of nitre-air, because when a candle is burnt in a closed vessel only a portion of the contained air is consumed. Nitre-air exists in large quantities in a condensed form in nitre; hence combustible bodies mixed with nitre will burn under water, or in a vacuum. The acid of nitre contains all the nitre-air in nitre, but it does not inflame bodies so readily as nitre, because in it the nitre-air is surrounded by particles of water which tend to quench the burning body. Nitre-air is not combustible itself, neither does nitre contain any combustible substance, because it may be fused in a red-hot crucible, but no ignition will be observed to take place until a combustible body has been added. All acids contain nitre-air: - how curiously this contrasts with Lavoisier's name oxygen, from οξέως γεννήτρια, which he gave to the gas, because he believed it to be an essential constituent of all acids. Sulfuric acid, according to Mayow, consists of nitre-air united with sulphur; wines become sour and are changed into vinegar by the absorption of nitre-air from the atmosphere. It is the cause also of fermentation and putrefaction, and for this reason, substances when covered with fat or oil do not putrefy. During calcination metals increase in weight, and this increase is attributed by Mayow to absorption of nitric air; thus calx of antimony is antimony plus nitre-air, and this is borne out by the fact that a substance absolutely similar to calx of antimony may be procured by treating the metal with the acid of nitre and evaporating. Again, rust of iron is iron united with nitre-air.

We come now to some of the first experiments in Pneumatic Chemistry. In one of his experiments Mayow supported a kind of ledge within a bell-jar full of air (see Fig. 20); upon the ledge he placed a piece of camphor, and fired it by concentrating the rays of the sun by a lens upon it. The camphor ignited and burnt for some time; water then rose in the jar; and on again attempting to ignite the camphor he was unsuccessful. A lighted candle was also burned in the jar with the same result. Thus a part only of the air had been consumed, and the remainder was unable to support combustion. The siphon tube (shown on the right-hand side of the figure) was inserted at the commencement in order to render the height of the water the outside the tube, as stoppered air-jars were then unknown.

FIG. 20.—Early experiment in pneumatic chemistry.

Pneumatic chemistry

Fig 20.-Early experiment in pneumatic chemistry.

Thus it was clearly proved that air is diminished in bulk by combustion. In order to prove that respiration produces a similar result, Mayo tied a piece of moist bladder over the mouth of a jar (Fig. 21) and upon this he pressed a cupping-glass, so that the edges fitted airtight. Within the cupping-glass he placed a mouse, and as the animal continued to breathe he noticed that the bladder was forced up more and more into the cupping-glass, proving that the air within it had been diminished by the respiration.

Fig. 21.—Early experiment in physiological chemistry.

Physiological chemistry

Fig. 21-Early experiment in physiological chemistry

Thus Mayow endeavored to establish a connection between combustion and respiration. He also placed a mouse in a vessel standing over water, and noticed that the water rose in the jar as the respiration continued; and he found it impossible to ignite a combustible body in a jar of air in which a mouse had been suffocated. Again, he placed a mouse and a lighted candle together in a jar of air, and he noticed that the mouse only lived half as long as a mouse lived in the same bulk of air without the candle. Air deprived of its nitre air was assumed to be lighter than nitre-air, because if a mouse is placed near the top of a closed vessel it dies sooner than if placed near the bottom.

In 1672 Robert Boyle procured hydrogen gas by acting upon iron filings with an acid, and proved its inflammability; but he does not appear to have further studied its properties, and its discovery is always attributed to Cavendish, a century later. Boyle suggests that it probably consists of "the volatile sulphur of Mars, or of metalline steams participating in a sulfurous nature." Mayow also procured some of this gas by acting upon iron with dilute sulfuric acid, and he proves that it is not a supporter of life.

Mayow's second treatise is on Respiration, and he herein expresses views far in advance of any of his predecessors. He proved that the nitre-air is alone concerned in respiration, and he asserts that this is absorbed by the blood, while the rest is rejected. It unites with combustible particles in the lungs, and thus produces animal heat The lungs consist of a number of minute sack-shaped membranes through which the nitre-air passes to the blood.

We add the following resume of Mayow's treatise, and of the position which it ought to occupy in the history of chemistry, from an article which we wrote on the subject a few years ago.

Mayow's work is remarkable in several respects. In it he conclusively proved that respiration and combustion are analogous processes; he upset the four-element theory by demonstrating the compound nature of air; and he recognized oxygen and nitrogen as clearly and almost as notably as they were recognized a hundred years later—the one the supporter of life and combustion, the principle of acidity, and the cause of fermentation and putrefaction, heavier than atmospheric air; the other incapable of supporting life or combustion, and lighter than atmospheric air. We find, moreover, in this work the dawn of the idea of chemical affinity in the fermentation, which he speaks of as taking place between nitre-air and combustible particles, and extending to the production or destruction of things. Mayow even employs some of the terms in general use in the present day; thus he speaks of affinity, existing between acids, and earthy substances, and uses the words combiner and combiner in speaking of the congresses of different substances.

The treatise is characterized by much clear and condensed thought, well-sustained argument, and accurate reasoning; moreover, we seldom meet with instances of too hasty generalization, always the dominant source of error in the early development of a science. We further observe a great advance towards that exact and discriminating mode of thought which is necessary for the investigation of chemical phenomena. The period in which Mayow wrote was, as regards chemistry, a period of transition; there was as yet no work on scientific chemistry, yet Mayow's treatise approached more nearly to such a work than that of any of his predecessors. The works of previous writers in this direction belonged to one of the three following classes: they were either chemical-metallurgic, chemical-medical, or alchemical treatises, or they partook of the nature of all three. The publication of works on alchemy was fast waning before the advances of the new philosophy; for as superstition retreats, and as men begin to devote their energies to the legitimate investigation of Nature, a false and chimerical art must of necessity cease to find votaries. Mayow was the first to discuss the intimate nature of an intangible body; other writers had treated of the air as a whole, but no one had endeavored to ascertain the nature of its internal constitution, or to determine why it produces certain changes in surrounding bodies, upon what these changes depend, and the nature of the constituent or constituents of the air producing them. The old dogma of the elemental nature of the air was received as an absolute truth, although entirely unproven; it was thought that a theory which had been received since the earliest ages must of necessity be correct, and no attempt was made to disprove it.

We see from the above that it was the investigation of the nature of nitre which led to the knowledge of the constitution of the air, and to the first experiments in Pneumatic Chemistry, Mayow remarks at the commencement of his treatise, that so much had been written about nitre, that it would appear "ut sal hoc admirabile non minus in philosophia, quam bello strepitus ederet; omniaque sonitu suo impleret;" and when we remember its connection with the foregoing results we are almost inclined to agree with him.

Chapter XI

The Theory of Phlogiston - Comparison with Hook's Theory of Combustion - Early Ideas regarding Calcination - Stephen Hales - His Pneumatic Experiments - Boerhaave - Conclusion.

About the year 1669 we find the first dawnings of a theory which was proposed in order to connect together various chemical phenomena, and notably for the explanation of combustion, the common and most obvious of all chemical actions. This theory, known as the "Theory of Phlogiston," powerfully influenced chemistry for a century; indeed, upon its ruins the structure of modern chemistry was raised by the labors of Lavoisier, Priestley, and Scheele. The proposers of this theory John Joachim Beccher (b. 1625, d. 1682) and George Ernest Stahl (b. 1660, d. 1734) endeavored to trace the cause of various phenomena, of chemical change to the assimilation or rejection of what they called "materia aut principium ignis non ipse ignis" -not actual fire, but the principle of fire; a something not much unlike the pure elemental, celestial fire which a few Ancient and many Middle Age writers had feigned to exist. Stahl believed this materia ignis to be a very subtle, invisible substance, which neither burns nor glows; its particles penetrate the most dense substances, and are agitated by a very rapid motion. When a body is burned it loses Phlogiston; when a body is unburned, if we may use such an expression, or deoxidised, it assimilates Phlogiston (φλογιστόν burnt). Thus if lead is heated for some length of time it is converted into a powdery substance which they called calx of lead, and we, lead oxide; the lead has lost Phlogiston, said Stahl. On the other hand, if this same calx of lead is heated with red-hot charcoal, it is de-oxidised and becomes lead again. It has now assimilated the Phlogiston, which it had before lost.

But here arose a difficulty. A metal was found to be heavier after calcination than before; thus loss of Phlogiston led to gain of weight, which was altogether anomalous, and apparently incapable of explanation. But the Phlogistians were equal to the occasion: the supporters of a pet theory will create any number of the most vague and impossible hypotheses rather than yield up their darling to destruction: so, said they, Phlogiston is a principle of levity; it confers negative weight, it makes bodies lighter, just as bladders attached to a swimmer lighten him.

The theory was applied as generally as possible:— thus sulfuric acid is produced by burning sulphur under certain conditions of oxidation; the sulphur loses Phlogiston, and becomes heavier, like the metallic "calx"; hence sulfuric acid is sulphur minus Phlogiston, while sulphur is consequently sulfuric acid plus Phlogiston. In fact, loss of Phlogiston was synonymous with what we call oxidation ; and gain of Phlogiston with de-oxidation. The existence of Phlogiston was so utterly unsupported - by experimental proof that the theory could scarcely exist without many opponents. The endurance of the most false and chimerical theory is often really wonderful. The Phlogistians were attacked first in one direction, then in another, yet the theory continued to find supporters. Finally, as a last resource, hydrogen gas—recently investigated by Cavendish - was said to be Phlogiston; but this was so entirely different from the Phlogiston of Stahl that the theory was now seen on all sides to be fast giving way. At length Lavoisier, a century ago, conclusively disproved the theory by means which cannot be discussed here, because they belong to the more advanced history of the science.

How the crude, unscientific, illogical theory of Phlogiston could have arisen in the face of Hooke's admirable Theory of Combustion and Mayow's experiments in support of it, must always remain a mystery. It is probable that if Mayow had not died a young man, or if Hooke had found leisure to prosecute his views, the Theory of Phlogiston would never have been propounded. The theory has been much over-praised. The only service which it rendered to the science was that it introduced a certain amount of order and system, which was hitherto wanting. It led to the grouping together of certain classes of facts, and, to a slight extent, to the application of similar modes of reasoning to similar chemical phenomena. And although that reasoning was altogether wrong, it seemed to indicate the means by which, with a more perfect and advanced system, chemistry might become an exact science subject to definite modes of treatment.

We have more than once spoken of calcination, which was indeed one of the most prominent operations of old chemistry. Since the examination of the process led to the proposal of just ideas concerning the materiality of the air — most often denied by Ancient and Middle Age writers — it may be well to glance at the early ideas regarding calcination. Here then was the dominant experiment in this direction : I take a bright lustrous metal, tin or lead, melt it, keep it in a molten state for a while, and it is converted into powder which weighs more than the original metal. Again I heat this same powder with charcoal, and it becomes metal again; yet nothing that can be seen has been added to the metal, or taken away from its calx. Geber defines calcination as "the pulverization of a thing by fire, by depriving it of the humidity which consolidates its parts." He observed that the metal increases in weight during the operation, although deprived of its humidity."

Cardanus asserted that the increase of weight in the case of lead amounted to one-thirteenth the weight of the metal calcined; and he accounted for it on the sup- position! that all things possess a certain kind of life — a celestial heat, which is destroyed during calcination; hence they become heavier, for the same reason that animals are heavier after death, for the celestial heat tends upwards. This idea was almost similar to that- of the Phlogistians; although published more than a century before Beccher wrote his Physica Subterranea. In 1629 Jean Rey, a physician of Bergerac, attempted >to discover the cause of increase, and attributed it to the absorption of "thickened air" (lair espessi) by the metal during calcination. Lemery, as we have seen, attributed the gain to the absorption of corpuscules de feu. Afterwards came the nitre air of Mayow; then a century later the increase was proved to be due to the union of the body with a constituent of the air which Lavoisier named oxygen gas; and this gas was first discovered by heating one of the calces (calx of mercury), about which so much speculation had been wasted, and so little experiment bestowed, by earlier writers.

We are drawing towards the end of our subject, but we think any account of the earlier history of chemistry would be very incomplete without a notice of the work of Dr. Stephen Hales (b. 1677 d.1761). In a number of Papers communicated to the Royal Society, and afterwards published in a work entitled Statically Essays we find a variety of experiments by Hales, chiefly relating to pneumatic chemistry. Herein there is an account of "A specimen of an attempt to Analyze the Air by a great variety of chymico-statical experiments, which show in how great a proportion air is wrought into the composition of animal, vegetable, and mineral substances, and withal how readily it resumes its former elastic state when, in the dissolution of those substances, it is disengaged from them." In order to determine the quantity of air disengaged from any substance during distillation or fusion, Hales placed the substance in a retort, and luted the retort to a large receiver with a small hole at the bottom; water was caused to occupy a known space in the receiver, and the amount of air expelled was estimated by noting the amount of water remaining in the receiver at the conclusion of the experiment, after cooling. Hales employed the apparatus (Fig. 22) to measure the volume of air generated by any kind of fermentation, also by the reaction of one body upon another.

Measuring a gas

Fig. 22 Hales method of measuring a gas.

Fig. 23 Measurement of the elastic force of the gas produced by fermenting peas.

The substances undergoing fermentation were placed in b, and over the whole a vessel, a y, was inverted, closed below by water in the vessel, x x, and containing above a certain amount of air, to the level, z. If air were generated, the water in a sank (say to y); while if air were absorbed by the bodies in b the water rose (say to n). Sometimes he placed different substances on pedestals in a jar of air, and ignited them, as Mayow had done, by a burning-glass, and noted the alteration in the bulk of air. He did this with phosphorus, brown paper dipped in nitre, sulphur, and other substances. If he required to act upon substances by means of a strong acid, he would place the substance in a suitable vessel on a pedestal in a known volume of air, standing over water, and would suspend over it a phial which could be emptied by pulling a string. These devices were closely copied by Priestley and Lavoisier in their experiments upon gaseous bodies. If a substance required to be heated violently, it was placed in a bent gun-barrel, r r (Fig. 24), one end of which was placed in a furnace, while the other was placed under a bell-jar, a b, full of water, inserted in the pail of water, x x. He distilled a number of substances, apparently taken at random, and determined the amount of gas evolved; but he appears to have been at no pains to determine the nature of the gas, assuming it to be ordinary atmospheric air. Thus he distilled one cubic inch of lard, and collected thirty-three cubic inches of gas as the product of decomposition. Tallow, horn, sale-ammoniac, oyster shells, peas, amber, camphor, and many other substances were similarly treated.

Two grains of phosphorus ignited in a closed vessel of air were found to absorb 28 cubic inches of air 211 grains of nitre mixed with bone ash yielded 90 cubic inches of gas ; 54 cubic inches of water on

Fig. 24.—Hales, pneumatic experiments.

Measuring elastic force

Fig. 24- Hales, pneumatic experiments.

boiling yielded 1 cubic inch of air. In order to measure the elastic force of the gas produced by fermenting peas, Hales filled a small, strong bottle, c (Fig. 23) with peas, filling up the interstices with water; mercury to a depth of half an inch was then poured in, and of course remained at the bottom of the vessel, c. A long tube, a z, the lower end of which dipped beneath the mercury, was securely fastened into the mouth of the bottle, b and fixed air-tight. In a few days' time the peas were in a state of fermentation, and the generated gas had forced the mercury to ascend in the tube a z to a height of 80 inches; hence the gas in c was existing under a pressure of about 35 lbs. on the square inch.

Hales also produced gases by various reactions. Thus he poured a cubic inch of sulfuric acid on half a cubic inch of iron filings: no effect took place until he had diluted the acid with water, when 43 cubic inches of air (as he calls it, in reality hydrogen gas) came off. Iron filings mixed with nitric acid, or with ammonia, or sulphur, were found to absorb air. A cubic inch of chalk treated with dilute sulfuric acid produced 31 cubic inches of air (in reality, carbonic anhydrous gas). If space permitted, we could say much more of Hales' work. His experiments on respiration, and on various principles of vegetation, are exceedingly ingenious, and often accurate. It has often been said that Lavoisier created modern chemistry by the introduction of the balance into chemical experiments; but here we find Hales weighing his substances, and measuring his gases, years before Lavoisier was born. Hales did not sufficiently, investigate the nature of the various gases which he produced in the course of his experiments, but he assuredly paved the way for many of the after-discoveries of Priestley, Cavendish, and Lavoisier.

Dr. Hermann Boerhaave, of Leyden (b. 1668, d. 1738), was a contemporary of Hales. He was the author of the first comprehensive system of chemistry: a bulky quarto in two volumes, entitled Elementa Chemiæ, which appeared in 1732, and which for many years was the chemical text-book if Europe. In it he defines chemistry as "an art which teaches the manner of performing certain physical operations, whereby bodies cognizable to the senses, or capable of being rendered cognizable, and of being contained in vessels, are so changed by means of proper instruments, as to produce certain determinate effects, and at the same time discover the causes thereof, for the service of various arts."

But, hold! our task was to give some account of the birth of chemistry, while a science with such a ponderous definition as the above is no longer infantile. The babe has grown up about us until it has assumed a tremendous individuality. The great discoveries of the fathers of modern chemistry, Lavoisier, Scheele, Priestley, Cavendish, Davy, need not be told here: they belong to the later history of chemistry. We have traced the science from its commencement in the crude metallurgical, and other operations of the ancients, to the time when a comprehensive system of chemistry appeared. When we think of the vast dimensions of the science of today, the numberless text-books which are to be found in every language, the great laboratories springing up in every country, the immense amount of original research, we are carried back in spirit to those mistaken, but often grandly energetic, men who said to the disciples of their art:

ORA!

LEGE, LEGE, LEGE, RELEGE, LABORA;

ET INVENIES.

THE END.

JUN 4 - 1913

LONDON; CLAY, SONS AND TAYLOR PRINTERS.

Lecture Notes

In Nature (Rodwell 1872) the author publish first chapters of the book (Rodwell 1874), with few organizational changes and late (Rodwell 1873) Mayow experiments. Accepting as proved by Boyle's experiments that air is necessary for combustion, Mayow showed that fire is supported not by the air as a whole but by a more active and subtle part of it. This part he called "spiritus igneo-aereus," or sometimes "nitro-aereus", for he identified it with one of the constituents of the acid portion of nitre . Chemical News of the era contain a serial of the book (Rodwell 1866), published (Rodwell 1874).

About the style of book,modern history cannot divine the past as prophets divined the future, obscure language of divination is the language of poetry. Taine's distinction between the important and the unimportant and his call for the elimination of confusion is from rhetoric. The pact between rhetoric and history is Machiavellian, and infinite for both, caught between the words and reality.(Lombardo 1990) Taine rejection of rationalism was a crucial first step in the elaboration of new forms of liberalism.(PITT 1998);

Time machine going a century back, (Boerhaave Herman 1727) notes that chemistry is growing up. For the name, found in index (Ada Adler 1939) send a helping hand, definition is "the art of...gold" In the book is discussed manuscripts, late translated (Earle Radcliffe Caley 2008) and receipts (Kreuzner 2011) and Democritus (Berryman 2013) and a short view of the word alchemia, the mystical concurrent.

In the Prolegomena or the history of chemistry, editions 1727 (Boerhaave Herman 1727) pg. XVII The first thing that occurs in our undertaking, is the word chemisty,...should be wrote... in Latin and English chemia and chemistry, not as usual Chimia and chymistry. From the 1st to the 4th centuries, alchemical practice develops itself into an art of metallic transmutation and two distinct alchemical "schools" seem to emerge: the one, represented by Ostanes, is still based on the practical knowledge of craftsmen, blacksmiths and dyers, although a shift is being accomplished from "chrysosis" (giving to a base metal the appearance of gold) to "chrysopoeia" (transforming a base metal to gold); the other, represented by Zosimos and Maria the Jewess, assumes a religious, Gnostic orientation, putting the emphasis on the elaboration of distillation techniques.

The period of Byzantium is a turning point, not only because there are many commentators of the ancient alchemical texts, but for the attempt, during the 10th century, to collect these texts and to articulate them in a coherent corpus, the surviving manuscript copies of which comprising, to our days, the main evidence for the emergence and the historical development of Greek alchemy...Kanaboutzes defines alchemy as τέχνη τῆς χυμίας ('art of "chymia"'). By the term χυμία (or any of its different spellings) Byzantines ascribed to what we call alchemy a certain philosophical and practical system. As his commentary is addressed to a Latin ruler, he shows interest in clarifying the etymology of the Latin-derived term ἀρκύμια ("archymia"), which is analyzed into "arte" and "chymia".

Accordingly, he defines the subject matter of the alchemical study in practical terms of the dissolution of metals.

Throughout the passage in question he repeats that χυμία is a mystical, secret and sacred. (Rampling 2012) The first author in whom the word is found, is Plutarh, who lived under the emperors Domitian, Nerva, and Traian That philosopher in his treat of Isis and Osiris, take the occasion to observe, that Egypt, in the sached dialect of the Country, was called as the black of the eye...chemia: by wich seems to intimate...

Late, (Boerhaave Herman 1753) pg. 7 - 8 opinion is another ...is so ancient, as perhaps has been used in the antediluvian age. On this opinion was Zosimos, the Panapolite, whose Greek writings, though known as before the George Agricolla, and afterwards pursued by...Scalinger and Olahus Boritius, still remain unpublished in the King of France's Lybrary... Zosimos...says expressly, that the art called chemia, was first taught by demons to the daughters of men, in consideration of certain favors. (Dayies 1868) pg 714

Many of the useful arts depend on chemical processes, in the sense in which we use the word "chemical" now. Metallurgy, dyeing, the manufacture of porcelain and glass, were all known in the early ages of the world, and are all chemistry in practice.

We find the name of Chemistry applied to the art of making gold and silver in the fifth century, in the earliest work known in which the word occurs. This work, entitled, A Faithful Description of the Sacred and Divine Art of making Gold and Silvery by Zosimos, the Panapolite, carries back the art to a far distant period, for it attributes it to the sons of God mentioned in the 6th chapter of Genesis, who, it states, were angels allured from heaven by the charms of women, to whom they imparted the secret of making precious metals. Suidas, in his Lexicon, written in the eleventh century, says, under the word chemia " The preparation of silver and gold.

The books on it were sought after by Diocletian, and burnt, on account of the new attempts made by the Egyptians against him.

It is, however, doubtful whether Alchemy can claim such high antiquity as this; for the silence of Latin authors, especially Pliny, on the subject, would lead us to believe that it took its rise among the Greeks at a later date. The earliest works on the subject are Greek, and a long list, comprising eighty works, is given in Boerhaave's Chemistry, 1753.

List of Boerhaave (Boerhaave Herman 1753) pg 18 - 20

1. Synesius; whose treatise of the philosopher's stone is in the library of Leyden.
2. Synesiuss the philosopher, on the book of Democritus, published entire in Fabricius (z).
3. Zosimus the great, the divine the Panapolite, of Panapolis, a city of Egypt, wrote XXIV books of imouth, or chemistry, to Theosebia, entitled Zosimus, the Panapolite, his faithful description of the sacred and divine art of making gold and silver; also of instruments and furnaces.
4. Olympiodorus, the Alexandrian.
5. Heliodorus, on the making of gold.
6. John , the high-priest, in the holy city, concerning the holy art.

7. Stephen, the universal philosopher, of Alexandria, his account of the holy and divine art of making gold: preserved in MS. in the library of Leiden
8. Orus, his chemical pieces
9. Sophar in Persia
10. Hermes, extant in sixth century, cited by Zosimus
11. Dioscourus, priest of the great Serapis, in Alexandria
12. Ostanes of Egypt, to Petasius concerning the sacre and divine art
13. Moses, the profet on chemical composition.
14. Mary, the Jew.
15. Pelagius, the philosopher on the holy and divine art.
16. Porphyry.
17. Epibychius, or Epibechius.
18. Comarius the philosopher and high priest his piece instructing Cleopatra in the divine and holy art of the phylosopher's stone.
19. Cleopatra, wife of king Ptolemey. The same on weight and measures.
20. Cosma, the monk, his interpretation of the art of making gold
21. Agathodsmon, his colle Hiou and comment on the oracle of Orpheus.
22. Pappus, the Philosopher, his work.
23. Heracleus, the king.
24. Salmana Arabus, his method.
25. Christianus, on holy water.
26. Theophrastus, the philosopher on the divine art.
27. Archelaus, the philosopher, on the divine art.
28. Claudian.
29. Sergius.
30. An anonymous philosopher, on chemistry
31. Michael Psellus on the making of gold. He lived under Constantine Duca, in the year of Christ 1060 (a).
32. Isis, the prophetess, to her son Orus.
33. J Blemmidas, his chemical work.
34. Nicephorus.
35. Democritus, his book to Leucippus.
36. Democritus, his physical and mystical writings,
37. Hierotheus the philosopher, concerning the philosopher's stone.
38. Isaac the monk; how to find the method of making silver (b),

(Thomson 1831) pg 6 "...The Scriptures inform us that the giants sprang from these embraces. Chema is the first of their traditions respecting these arts. The book itself they called Chema; hence the an is called Chemia" Zosimuss is not the only Greek writer on Chemistry. Olaus Borrichius has given us a list of thirty-eight treatises, which he says exist in the libraries of Rome, Venice, and Paris: and

Dr. Shaw has increased this list to eighty-nine. But among these we find the names of Hermes, Isis, Horus, Democritus, Cleopatra, Porphyry, Plato, &c.- names which undoubtedly have been affixed to the writings of comparatively modern and obscure authors. The style of these authors, as Borrichius informs us, is barbarous. They are chiefly the production of ecclesiastics, who lived between the fifth and twelfth centuries. In these tracts, the art of which they treat is sometimes called chemistry; sometimes the chemical art sometimes the holy art; and the philosopher's stone.

It is evident from this, that between the fifth century and the taking of Constantinople in the fifteenth century, the Greeks believed in the possibility of making gold and silver artificially; and that the art which professed to teach these processes was called by them Chemistry.

These opinions passed from the Greeks to the Arabians...

(Berthelot 1885) late, in 1885 will extend bibliographical research. Here we found the oldest manuscript on alchemy, Leiden manuscript translated late (Earle Radcliffe Caley 2008)

The Greeks invented eponymi and included them in the heroic genealogies because they wished to account for present realities.

The etiologist started with an awareness of such realities as the Apis bull or the city of Nineveh and furnished for them an explanation based on probability. He sought no new information, but attempted to order and schematic the information which he already had.(ROBERT DREWS 1973)

Scientific works of Democritus are sadly known only from second hand resources. (Chitwood 2004) Many consider the best introduction to the problem of fragments still to be found in the discussion in P. Natrop, Die Ethica des Democritos, Text and Untersuchungen (Wouters, Kolar 2011)

(Christopher Charles Whiston Taylor 1998) give as an compendium of sources for Democritus and Leucippus and problem of first editions of works remain (Fergurson 1894) "The tradition according to which, among other knowledge, the art of ennobling metals had been brought from heaven to earth by demons, was universally diffused in the first centuries of our era; Zosimos of Panopolisstates that the mystical book from which this art was to be learned was termed, chemia and the art itself chemistry. This myth doubles sprang from one exactly similar which is to be found in the apocryphal book of Enoch; indeed indications of it are to be met with even in Genesis. The later alchemists were inclined to refer the origin of alchemy to the time before the flood,thinking that a special sanctity would accrue to their art from this great age.
Moreover, they wrote down various biblical characters as alchemists, on the authority of certain passages in Holy Writ, for instance, Moses and his sister Miriam, and the Evangelist John. When legends such as these found credence in the Middle Ages,it is hardly surprising that the records as to the origin of this art, which remain to us from ancient times, should have upheld their authority over a very long period." (Meyer 1906)

Chapter two between the mythology , (Christopher Charles Whiston Taylor 1998) Egypt India and China , Leukippos will introduce indivisible atoms and his pupil Democritus an empty space between they. A pre Socratic reader (Curd, McKirahan 2011) remember the religion contributions. Aristotle have a special part here and is support for a jump of more than 1500 years to Pretiosa Margarita (Lacinio, Boni 1546). Wandering in neighborhood in search of eternal life first scientific experiments are described (Porto 2002) in charge here is an physician. That's an excellent opportunity for connections (Mahdihassan 1979) , (Mahdihassan S. 1981)

The "Art of making gold" is back in Chapter III, illustrated: (Wilkinson 1871) At the end shield of Achilles is is analyzed.

Processes and receipts, the core of chemistry, are slow moved to paintings (Davy 1815) with connections of green glass at Greeks with cobalt, but what if the Smaragdine table is containing uranium?

Chapter five will have a short introduction of the symbols and at six we are back to alchemy, Suidas, (Curd, McKirahan 2011) is an revision of all philosopher above. Zosimos still have surprises (El Khadem 1995) and (Berthelot 1885) is a strong supporter, we a getting close to that point where Casiopeea of Cleopatra will disappear (van Tiggelen 2008)

Will meet Hermes Trismegistus, and Olympiodorus (Wildberg 2013) but is missing Erasmus (Jensen 2006) and Casiopeea of Cleopatra, an alembic for Aqua Vitae is missing. Roger Bacon have a central position for English manuscripts (Klein 2013), late will learn that Secret Secretorum have another name, The Mirror of Alchemy (Bacon, Davis 1931) and in 1590 is printed (Bacon, Linden 1992) and electronic format (Bacon, Roger, 1214?-1294 1597) All of them a present in some books with Geber (Geber et al. 1545) in Dresden or Vatican (Geber 1542). (STEELE 1931) give us another catalog of manuscripts and a nice review.

"Chemical technology is purely Byzantine and Hellenic, but it was altogether divorced from theory, and the "Turba" is never quoted by any writer in the first century and a half of alchemical literature. But though Greek thought had no direct action on Latin writers, it was the ultimate source of Arabic alchemy as regards the theory of metals, the elixir of life being apparently of Chinese origin. No doubt a few Byzantine adepts found their way into the west? Roger Bacon mentions a Greek he had known, and there is the still earlier case of the Jew of Bremen. The story of translation from the Arabic begins with Robert of Chester in 1144, Plato of Tivoli and Hugo Sanctallensis, and goes on to Gerard of Cremona, who before his death in 1187 translated three classics of alchemy? two of them only printed in our own time, by Berthelot and the writer. A number of treatises were translated before the middle of the thirteenth century, as shown by the quotations from them in the "Speculum Naturale" of Vincent of Beauvais (1245), the alchemical writings of Roger Bacon ending 1267, and the writings of Albertus Magnus on minerals and the "Speculum Astronomicum". St. Thomas also accepts the scientific possibility of alchemy"

Raymond Lulli aka Ramon Lull "Doctor Illuminatus", apparently have second hand manuscripts after unknown manuscripts.

Chapter IX give us an abstract with images from (Giano Lacinio et al. 1894) and in X is born. Paracelsus split Iatrochemy from his own metaphysical world, and at XI (Boerhaave Herman 1727) give inorganic and organic chemistry. (Boerhaave Herman 1753a), (Boerhaave Herman 1753b)

Publication bibliography

Ada Adler (1939): Suidae Lexicon II-V (Lexicographi Graeci I). In The Journal of Hellenic Studies 59, pp. 176–178. Available online at

uri:jstor.org/stable/626979.

Bacon, Roger; Davis, Tenney L. (1931): The mirror of alchemy of Roger Bacon. In J. Chem. Educ. 8 (10), p. 1945. DOI: 10.1021/ed008p1945.

Bacon, Roger; Linden, Stanton J. (1992): The mirror of alchimy. Composed by the thrice-famous and learned fryer, Roger Bachon. With certaine other worthie Treatises of. New York: Garland Pub (Garland reference library of the humanities, 1031). Available online at

http://goo.gl/BtJ7iA.

Bacon, Roger, 1214?-1294 (1597): The mirror of alchimy., composed by the thrice-famous and learned fryer, Roger Bachon,sometimes fellow of Martin in Oxenforde. Also a most excellent and learned discourse of the admirable Colledge: and force and efficacie of art and nature, written by the same author. With certain afterwards of Brasen-nose Colledge. With certain other worthies Treatises of: London :: Printed [by Thomas Creede] for Richard Oliue; Ann Arbor, Michigan: University of Michigan, Digital Library Production Service. Available online at

http://goo.gl/Xs2z46.

Berryman, Sylvia (2013): Democritus. First published Sun Aug 15, 2004; substantive revision Wed Aug 25, 2010. In Edward N. Zalta (Ed.): The Stanford Encyclopedia of Philosophy. Winter 2013. Available online at

http://goo.gl/8HY4uk.

Berthelot, M. (1885): Les origines de l'alchimie. Paris: G. Steinheil. Available online at

http://goo.gl/mM9cDM.

Boerhaave Herman (1727): A new method of chemistry. Open Library. xvi, 383, [1], 160, [161]-[188], 161-335, [45] p., 2 leaves of plates : With assistance of P. Shaw & E. Chambers. Fisher copy: Armorial binding with the crest of James Stuart on upper cover, and his monogram on lower. London: J. Osborn and T. Longman. Available online at

http://goo.gl/p15zsJ, checked on 8/20/2014.

Boerhaave Herman (1753a): A new method of chemistry. including the history, theory, and practice of the art. translated from the original Latin of Dr. Boerhaave's Elementa chemiae, as published by himself. With assistance of Peter translator Shaw. The third edition corected. 1 volume. London: T. and T. Longman (2). Available online at

http://goo.gl/6zQEvw, checked on 8/21/2014.

Boerhaave Herman (1753b): A new method of chemistry: including the history, theory, and practice of the art: translated from the original Latin of Dr. Boerhaave's Elementa chemiae, as published by himself. The third edition corrected. 2 volumes. London: T. and T. Longman (2). Available on-line at http://goo.gl/74W0PL, checked on 8/21/2014.

Chitwood, Ava (2004): Death by philosophy. The biographical tradition in the life and death of the archaic philosophers Empedocles, Heraclitus, and Democritus. Ann Arbor: University of Michigan Press.

Christopher Charles Whiston Taylor (1998): The Atomists. Leucippus and Democritus. Fragments a text and translation. In Journal of the Classical Association of Canada XXXVI/Phoenics Presocratics (Volume V). Available online at

uri:jstor.org/stable/41664274, checked on 8/18/2014.

Curd, Patricia; McKirahan, Richard D. (2011): A Presocratics reader. Selected fragments and testimonia. 2nd ed. Indianapolis: Hackett Pub. Co.

Davy, Humphry (1815): Some experiments and observations. on the colours used in painting by the Ancients. In Philosophical Transactions II. Available online at

http://goo.gl/UnGJhE

Dayies, E. (1868): Alchemy. In Liverpool, Literary And Philosophical Society Of. Available online at:

http://goo.gl/PBhhvG, checked on 8/25/2014.

Earle Radcliffe Caley (2008): The Leiden and Stockholm Papyri. Greco-Egyptian Chemical Documents From the Early 4th Century AD. With assistance of William B. Jensen. Cincinnati, OH (The Oesper Collections). Available online at

http://goo.gl/RR61r1, checked on 8/14/2014.

El Khadem, Hassan S. (1995): A Lost Text By Zosimos Reproduced in an Old Alchemy Book. In Journal of Chemical Education (ACS Publications and Division of Chemical Education). Available online at DOI: 10.1021/ed072p774, checked on 8/27/2014.

Fergurson, John (1894): On the first edition of the chemical writings of Democritus and Synesius. November, 1884 - October, 1894. Read before the Society, 19th November 1884. 32 volumes. Glasgow: Robert Andersen (Procedings, Philosophical Society of Glasgow). Available online at

http://goo.gl/5LV9oK checked on 8/14/2014.

Geber (1542): Geberis philosophi perspicacissimi. summa perfectionis magisterii in sua natura. ex Bibliothecae Vaticanae. With assistance of Kalid Rashaidibi, Avicenae, Merlini. Venetiis: Apud Petrum Schoeffer. Available online at

http://goo.gl/HAOMEm

Geber; Bacon, Rogerus; Richardus; Kalid; Hermes; Hortulanus (1545): Alchemiae Gebri Arabis philosophi solertissimi, Libri. Cum Reliquis, ut uersa pagella indicabit. Ioan[nes] Petreius Nurembergen[sis] denuo Bernae excudi faciebat. [Norimbergae]: Petreius (Bibliotheca Palatina, H1194).

Giano Lacinio; Arthur Edward Waite; Pietro Antonio Boni (1894): The new pearl of great price. A treatise concerning the treasure and most precious stone of the philosophers. Or the method and procedure of this divine art ;. with observations drawn from the works of Arnoldus, Raymondus,

Rhasis, Albertus, and Michael Scotus (1894). The original Aldine edition translated in English: London : J. Elliott and co.

Jensen, William B. (2006): Erasmus on Alchemy. In Bull. Hist. Chem., 31 (2), pp. 58–64.

Klein, Jürgen (2013): Francis Bacon. In Edward N. Zalta (Ed.): The Stanford Encyclopedia of Philosophy. Winter 2013.

Kreuzner, Christina (2011): Alkanna tinctoria (L.) Tausch as Purple Dye. in the Recipes of Papyrus Holmiensis and Papyrus Leidensis X. In Jan Wouters, Jana Kolar (Eds.): e-Preservation. Science. scientific research for the preservation of cultural heritage, pp. 123–130. Available online at

http://goo.gl/twgMxM

Lacinio, Giano; Boni, Pietro Antonio (1546): Pretiosa margarita nouella de thesauro. Ac pretiosissimo philosophorum lapide. Artis huius diuinae typus, & methodus: collectanea ex Arnaldo, Rhaymundo, Rhasi, Alberto, & Michaele Scoto;. Venetiis: Apud Aldi filios. Available online at:

http://goo.gl/5aOvyB

Lombardo, Patrizia (1990): Hippolite Taine between Art and Science. Reading the Archive: On texts and Institutions (1990), pp 117 - 133. In Yale French Studies 77, p. 117.

Mahdihassan, S. (1979): A comparative study of Greek and Chinese alchemy. In Am J Chin Med 7 (2), pp. 171–181.

Mahdihassan S. (1981): Alchemy and its fundamental terms. in Greek, Arabic, Sanscrit and Chinese. In Indian Journal of History of Science 16 (1), pp. 61–76. Available on line at

http://goo.gl/l6c8oW

Meyer, Ernst von (1906): A history of chemistry. from earliest time to present day. also an introduction to the study of the science (1906). With assistance of George McGowan. Third after third German edition with alterations. London: Macmillan. Available on line at

http://goo.gl/X6Yd5T checked on 8/20/2014.

PITT, ALAN (1998): The Irrationalist Liberalism Of Hippolyte Taine. In The Historical Journal 41 (04), pp. 1035–1053. Available online at

http://goo.gl/VTMdLO

Porto, Paulo A. (2002): "Summus atque felicissimus salium". The Medical Relevance of the Liquor Alkahest. In Bulletin of the History of Medicine 76 (1), pp. 1–29. DOI: 10.1353/bhm.2002.0038.

Rampling, Jennifer (2012): Byzantine and post-Byzantine Alchemy: Principles, Influences and Effects. With assistance of Rémi Franckowiak. In Gianna Katsiampoura (Ed.): Scientific cosmopolitanism and local cultures. Religions, ideologies. Societies. Book of Abstracts. 5th International Conference of the European Society for the History of Science. Athens, 1 - 3 November. Institute of Historical Research/National Hellenic Research Foundation

ROBERT DREWS (1973): The Greek Accounts of Eastern History. CAMBRIDGE, MASSACHUSETTS: The Center for Hellenic Studies, Washington, D.C.

Rodwell, G. F. (1872): Nature. A Weekly Illustrated Journal of Science. Volume 6, 1872 October 10, [No. 154]. The Birth of Chemistry: Macmillan and Co.

Rodwell, G. F. (1873): The Birth of Chemistry. IX. In Nature iII (April 24), pp. 492–494. Available on line at

http://goo.gl/r37dHC

Rodwell, G. F. (1874): The birth of chemistry. London: Macmillan (Nature series). Rodwell, George F. (1866): On the Supposed Nature of Air. prior to the discovery of Oxygen. In Chemical News and Journal of Industrial Science (348), pp. 51–53. Available online at

http://goo.gl/DvFSCs, checked on 9/1/2014.

STEELE, ROBERT (1931): Union Académique Internationale Catalogue. of Latin and Vernacular Alchemical Manuscripts. in Great Britain and Ireland dating from before the XVI Century. In Nature 127 (3192), pp. 6–7. DOI: 10.1038/127006a0.

Thomson, Thomas (1831): The history of chemistry. vol 2. 2 volumes. London: Henry Colburn and Richard Bentley (2). Available on line at

http://goo.gl/EVM9q1

van Tiggelen, Brigitte (2008): Neighbours and territories. The evolving identity of chemistry : 6th International Conference on the History of Chemistry : proceedings. With assistance of José Ramón Bertomeu-Sánchez, Duncan Thorburn Burns. Leuven: Mémosciences asbl. Available on line at

http://goo.gl/TUrU1i

Wildberg, Christian (2013): Olympiodorus. In Edward N. Zalta (Ed.): The Stanford Encyclopedia of Philosophy. Winter 2013. Available on line at

http://goo.gl/XH9Tus

Wilkinson, John Gardner (1871): A popular account of the ancient Egyptians. vol 2. 2 volumes. New York: Harper & brothers; Ann Arbor, Michigan: University of Michigan Library (Making of America (MOA)).

Wouters, Jan; Kolar, Jana (Eds.) (2011): e-Preservation. Science. scientific research for the preservation of cultural heritage. Morana RTD. Available on line at

http://goo.gl/QNbGyF

www.ingramcontent.com/pod-product-compliance
Lightning Source LLC
Chambersburg PA
CBHW081142170526
45165CB00008B/2767